Light Pollution, Urbanization and Ecology

Edited by Levente Hufnagel

Published in London, United Kingdom

IntechOpen

Supporting open minds since 2005

Light Pollution, Urbanization and Ecology
http://dx.doi.org/10.5772/intechopen.87456
Edited by Levente Hufnagel

Contributors
Agustinus Gunawan Admiranto, Rhorom Priyatikanto, Siti Maryam, Elyyani Elyyani, Siti Kurniawati, Muhammad Faizal Eko Saputro, Redhwan Al-Naggar, Lutfi Al-Maktari, Rajesh Singhal, Rekha Sodani, Udit Nandan Mishra, Subhash Chand, Indu, Hirdayesh Anuragi, Kailash Chandra, Jyoti Chauhan, Bandana Bose, Vivek Kumar, Gopal Shankar Singh, Devidutta Lenka, Mohd Kher Bin Hussein, Nor Syuriaty Bt Jaafar, Levente Hufnagel, Ferenc Mics, László Nowinszky, Lionel Hill, János Puskás, Károly Tar

First published in London, United Kingdom, 2022 by IntechOpen
IntechOpen is the global imprint of INTECHOPEN LIMITED, registered in England and Wales, registration number: 11086078, 5 Princes Gate Court, London, SW7 2QJ, United Kingdom
Printed in Croatia

British Library Cataloguing-in-Publication Data
A catalogue record for this book is available from the British Library

Additional hard and PDF copies can be obtained from orders@intechopen.com

Light Pollution, Urbanization and Ecology
Edited by Levente Hufnagel
p. cm.
Print ISBN 978-1-83880-972-0
Online ISBN 978-1-83880-973-7
eBook (PDF) ISBN 978-1-83880-974-4

For EU product safety concerns:
IN TECH d.o.o., Prolaz Marije Krucifikse Kozulić 3, 51000 Rijeka, Croatia, info@intechopen.com or visit our website at intechopen.com.

Meet the editor

Dr. Levente Hufnagel is an associate professor and the head of the Research Institute of Multidisciplinary Ecotheology, John Wesley Theological College, Budapest, Hungary, working on ecology, biogeography, ecological research methodology, and sustainability. He has more than 20 years of experience in leading Hungarian academic institutions teaching Ph.D., MSc, and BSc students from various social and cultural backgrounds. He has more than 300 scientific publications (in both aquatic and terrestrial ecological aspects of plants, animals, and microbes at both the community and population levels) and more than 1100 independent citations to his credit. As a participant in several big ecological research and development projects, Dr. Hufnagel has significant experience in multidisciplinary collaborations (with more than 200 coauthors in different publications). He has supervised several Ph.D., BSc, and MSc theses, and served as editor in chief of an international scientific journal. Dr. Hufnagel graduated from Eötvös Lorand University with a master's degree in Ecology and Evolutionary Biology and a Ph.D. in Hydrobiology. He also has a Ph.D. in Agricultural Science from Szent István University, Hungary, and several other degrees from the Corvinus University of Budapest and Adventist Theological College.

Contents

Preface

A small but well-defined part of the global ecological crisis of our time is what we call the phenomenon of urbanization, an important part of which is light pollution.

Sunlight is one of the most important abiotic factors that regulate the development of almost every lifeform. During evolution, humans and animals developed many receptors for the perception of light, which helps living things obtain information about the amount, quality, seasonal changes, direction, and length of illumination. During the dark period, regeneration processes begin in cells and organs to repair the damage that occurred during the day. Urban lighting makes it harder for living beings to repair damage caused by oxidative stress. In the 20th century, the size of the areas illuminated by artificial light and the intensity of the light increased rapidly, as did its impact on wild and synanthropic species and humans.

This book presents a comprehensive review of light pollution. Following a comprehensive ecological introduction, chapters examine the effects of light pollution on plants, invertebrates, and humans using case studies, discuss the trend of increasing environmental impact, and finally, highlight eco-architecture as a possible solution in the visual symbiosis between the biosphere and humankind.

Light Pollution, Urbanization and Ecology provides interesting information about the frontiers of this scientific area. It is a useful resource for anyone interested in the problems of light pollution ecology.

Levente Hufnagel
Research Institute of Multidisciplinary Ecotheology,
John Wesley Theological College,
Budapest, Hungary

Chapter 1

Introductory Chapter: Ecological Effects of Light Pollution - A Review

Levente Hufnagel and Ferenc Mics

1. Introduction

Ecologists have long studied the regulatory role of natural light in species interactions, but the effect of overnight electrical lighting was only recently studied. In the 20th century, the size of the area illuminated by artificial light and the intensity of the light increased rapidly, as did its impact on wild species, however, light trapping has also become important in scientific research [1–4]. A distinction is made between astronomical light pollution, which obscures a clear night sky, and ecological light pollution, which modifies the natural light conditions of the environment. Adverse effects have been observed in a number of taxonomic groups, such as the extinction of migratory birds due to high, illuminated structures, or the orientation of hatching sea turtles is also disturbed by the light of surrounding settlements. The impact of artificial night light on animals is a new challenge for researchers and conservationists. Mankind has been looking for opportunities since prehistoric times to make the days longer and to be able to orientate themselves at night. In the pre-industrial era, they were able to be active in the dark by lighting wood or oil. The impact of this on wildlife was still very limited. After the discovery and spread of the electricity, it has changed natural day and night light conditions in an increasing part of the planet. The term light pollution has been used for years, but not in an ecological sense, but in the sense that celestial bodies are harder to observe because of the light from the lamps. The combined effect of many millions of terrestrial light sources causes the "skyglow" phenomenon, making the stars less visible to the human eye. We will continue to addres light pollution in ecological term, which affects the behavior of living things. Large areas are lit unnecessarily, which is a great waste. In the US alone, 3600 billion kilowatt-hours of electricity was produced in worth of 362 billion USD in 2018 [5]. About 30% of outdoor lighting is a waste because it is turned on even when it is not needed or is not directed to where it needs to be [6]. Verheijen [7] wrote about this phenomenon as "photopollution" first, suggesting that artificial light has a detrimental effect on wildlife. This embraces unusual light intensities and unexpected fluctuations. The source of light pollution is the phenomenon of "nightglow", illuminated buildings, towers, street lighting, vehicles, oil rigs. They are all, more or less, disrupting the normal functioning of ecosystems, almost all over the planet today [8]. This harmful effect of anthropogenic activity on the ecosystem occurs almost everywhere on the inhabited continents, the only exception is Antarctica [9]. The highest light pollution is observed in temperate forests and the Mediterranean climate zones. The effect is least intensive in the Arctic tundra and alpine areas [10]. In the tropics, wildlife is particularly severely affected, as the length of the illuminated period varies little

IntechOpen

during the year [11]. In tropical zone, the species have evolved over the course of evolution to vary the length of days only minimally seasonally. Species living in the temperate zone have adapted to seasonal changes in the length of the day, which is also disturbed by artificial lighting and, through this, the behavior of the animals. At full moon the illumination is between 0.1–0.03 cd/m^2 and at new moon it is between 0.01–0.003 cd/m^2, in case there is no light pollution [12]. Significant deviation from this values affects the physiology and behavior of both plants and animals, as well as the functioning of the entire ecosystem.

2. Light and ecology

Measuring the intensity and cycle of natural light is the first step in being able to determine the extent of pollution at a given location. The intensity, spectrum and period of the light may vary. Ecologists ideally measure the amount of photons per unit area per second. More often, however, light intensity is measured in lux, which expresses the intensity of light perceived by the human eye. In this case, the visible light to the human eye is measured primarily and less emphasis is placed on the invisible range. Since many animal can perceive light of a different wavelength than humans, it is worth measuring a wider range, because wavelengths out of range sensible to human eye can affect affect the behavior of animals [13]. The instruments are designated to be most sensitive in range that can be perceived by human eye, which makes the measurement more complicated [14]. For example, a light source may emit UV light that is not perceptible to the human eye, but some animals are attracted to UV source [15]. Orientation and disorientation are responses to changes in external lighting. Attraction and repulsion, on the other hand, are a response to the light source itself and its luminance [16, 17]. Increased illumination prolongs the activity of species active during the day, or at dusk, because they are still able to orientate. Many diurnal species continue to forage by artificial light [18]. Therefore, the so-called "night light niche" have developed in cities, which is very beneficial to certain species as they can seek food for longer periods of time. This is, of course, detrimental to their prey species [19]. In the case of birds, territorial behavior, i.e. singing, also takes longer due to artificial light [20]. Continuous artificial lighting, however, causes disorientation in nocturnal animals. The best-known example is the disrupted migration of sea turtle hatchlings to the water They normally move away from the low, dark silhouttes, which is generally the surrounding vegetation, to reach the water as quickly as possible. Due to the light of the surrounding settlements, the plants have no shade or the direction of the shade changes. This confuses the little turtles and they loose direction. Females, preparing to lay eggs, can also be disturbed by the light from the villages [21]. Nocturnal animals have adapted to low light intensity, and they can be blinded by sudden strong light, making them disoriented. In some cases, the adaptation of the eye can take up to hours, during which the animal loses its ability to orient itself [22]. In the case of birds, the light source sometimes traps the animals, they do not leave the surrounds of illuminated structures. Birds may collide with each other, the lit-up building, or fall prey to predators [23]. In the case of insects, many taxa are known to be attracted to night light. Such as moths, veils, beetles, bugs, quails, mosquitoes, flies, buzzards, wasps and crickets. Attraction depends on the spectrum of light emitted [24]. In the case of non-flying arthropods, the reaction can be varied. Nocturnal species are often repelled by night light, trying to move away from it. Others prefer to take advantage of artificial light [25]. Some insects always show positive phototaxis as adaptive behavior, while others always show negative [26, 27]. In some cases, phototaxis can be exploited to protect animals. For example, in the case of the American eel (*A. rostrata*), animals can be diverted to a detour route if a dam or power plant on the river prevents the migration [28].

Artificial light also affects reproduction. The calling sound of male frogs (*Physalaemus pustulosus*) changes around urban cities without increasing the risk of predation, as there are fewer predators due to light. For this reason, their reproduction is also more successful than in the case of individuals living far from the city [29]. The reproductive success of birds is also affected by artificial lighting, but here the effect is more negative. The brooding birds are disturbed in their sleep, resulting in a significant reduction in reproductive success [30]. Visual communication, both within and between species, is strongly influenced by the light of the lamps. Some animals use the chemical reaction of bioluminescence for communication. The best known example of this is the fireflies (Coleoptera: Lampyridae). Due to the unnatural light intensity, their signals are less noticeable, so they are less able to make contact with their peers, and their chances of reproduction also deteriorate [31]. The coyote (*Canis latrans*) holds the team together at night with vocals and also signals the boundary of the territory with its voice. Due to the light of the settlements, the pattern of vocalization changes, it is most pronounced in the new moon, when it is otherwise darkest at night [32]. Vision plays a very important role in orientation, no wonder the changed light conditions also affect the behavior of the animals. This should also be taken into account in nature conservation, as some species may decline as a result of light pollution. For other species, the effect is positive, but from the point of view of community ecology, this effect may not be positive. The connection between the *Physalaemus pustulosus* frog and Corethrella spp. parasite midges is also affected by the lights of the lamps. The parasite is very sensitive to light pollution, so significantly fewer parasites can be observed near the settlements. The relationship between the parasite and the host species is further complicated by the fact that noise pollution also affects the abundance of the parasite [33]. A frog called *Hyla squirrela* feeds in the range of 10–5 lux and ceases to feed above 10–3 lux [34]. *Bufo boreas* feeds between 10 and 1 and 10–5 lux, while *Ascaphus truei* only feeds below 10–5 lux [35]. These species are not sympatric, they live in same area, niche separation occurs along the gradient of illumination, hence interspecific competition increases under the influence of artificial light. It is a well-known fact that the light of street lights attracts flying insects and insects attract bats [36]. But not all bats are attracted to light. How attractive a particular species finds the abundance of food provided by a lamp depends primarily on how fast it can fly. Slower-flying species are less likely to come to illuminated places because they have to compete with faster species there, and despite the increased food density, they would not be able to take more prey. Therefore, they prefer to avoid light and hunt in the dark [37]. So in competing communities, changed light conditions also change interspecific relationships. Some species benefit from a longer clear period because they can search for food for a longer period of time, but this also has disadvantages because predation risk also increases during food search [38]. Strong lighting makes rats more cautious, spends little time in the illuminated area, and their movement is also slowed near lamps, and the time spent searching for food is shortened [39]. During a full moon, the predation of marine zooplankton by fish increases because plankton congregate near the surface due to illumination. Due to artificial lighting, this event has become commonplace, greatly increasing the predation pressure from which fish benefit [40]. Yurk and Trites [41] observed that the hunting success of seals (*Phoca vitulina*) is significantly enhanced by coastal lights. So at the next trophic level, the proximity of settlements also increases the efficiency of hunting. But there is an counter-example. Indian peafowl (*Pavo cristatus*) becomes more alert and sleeps less when exposed to artificial light. This just reduces predation risk [42]. Species belonging to the families Geometridae,

Noctuidae, and Notodontidae families have sensory organs that detect ultrasound and try to evade bats. Conversly, by the light of mercury vapor lamp they does not exhibit evasive behavior. They do not seem to respond to ultrasound in the area lit by mercury vapor lamp. They do not also react to ultrasound flying in natural daylight. They detect ultrasound during the day, but then the source of the sound does not pose a threat to them, because in light they mostly perceive the sound emitted by Orthoptera species, which does not pose a threat to them. The light of mercury vapor lamp is interpreted by the animals 'nervous system as natural daylight and there is no danger of predation by bats. Therefore, they do not even try to avoid the source of ultrasound, significantly increasing the mortality rate and bat hunting success. It appears that in the light of led lamps that do not emit UV light, the evasive reaction is largely retained, and in the case of led lamp light, the power-dive maneuver is more often observed [43]. Altered animal behavior, as a spillover effect, endangers the functioning of the entire ecosystem in the long run. Good example is the diurnal drift of zooplankton. Zooplankton migrates vertically in the water body in a daily rhythm. They come close to the surface at night to feed and also follow the lunar cycle. Thus, they can avoid predation risk by fish [44]. Luarte et al. [45] during the study of zooplankton and invertebrate species found that artificial lighting reduced migration, and thus the time spent feeding. As a result, the size of the population has also decreased. Due to the decrease in predation pressure, the biomass of algae increased. Returning to the bats, the species that can fly slower do not gain anything from the food abundance caused by the light of the streetlamps due to the interspecific competition with faster species (*Lasiurus* spp., *Eptesicus* spp., *Nyctalus* spp. and *Pipistrellus* spp.) Therefore they prefer to continue hunting in the dark (*Rhinolophus* spp. *Plecotus* spp. and *Myotis* spp.) This leads to changes in community structure [46].

Sunlight is one of the most important abiotic factors that regulates plant development. During evolution, they developed a wide variety of receptors for the perception of light, which are used to obtain information about the amount, quality, seasonal changes, direction, length of illumination. Luarte et al. [45] during the study of zooplankton and invertebrate species found that artificial lighting reduced migration, and thus the time spent feeding. As a result, the size of the population has also decreased. Due to the decrease in predation pressure, the biomass of algae increased. Returning to the bats, the species that can fly slower do not gain anything from the food abundance caused by the light of the streetlamps due to the interspecific competition with faster species (*Lasiurus* spp., *Eptesicus* spp., *Nyctalus* spp. and *Pipistrellus* spp.) Therefore they prefer to continue hunting in the dark (*Rhinolophus* sp. *Plecotus* spp. and *Myotis* spp.) This leads to changes in community structure [46].

Sunlight is one of the most important abiotic factors that regulate plant development. During the evolution, many receptors have developed in them for the perception of light, with the help of which they obtain information about the amount, quality, seasonal changes, direction and length of illumination. This allows plants to adjust their biological clock and adapt life processes to optimal conditions. The mechanism that regulates the circadian rhythm is conserved and found almost throughout the kingdoms [47]. Plants utilize the energy of light during photosynthesis to produce organic matter. A wide variety of physiological processes, both light and dark, are required to coordinate regulatory processes. Light directs various plant physiological processes from seed germination to seed maturation [48, 49]. Changes in daily, monthly, and annual cycles may result in a shift in phenological phases caused by artificial light. A good example of this is the study of Škvareninová et al. [50] who observed sycamore (*Acer pseudoplatanus*) and stag's horn sumach (*Rhus typhina*) individuals in a park of Zvolen (Slovakia) between 2013 and 2016. On average, the autumn phenophase began 13–22 days later and lasted 6–9 days longer due to the light from the street lamps. As a result, however, they are more exposed to frost damage in late fall. French-Constant et al. [51] examined the phenology of urban trees' spring bud bursting in the UK. The trees around the lamps (*Acer psuedoplatanus*, *Fagus sylvatica*, *Quercus robur*, *Fraxinus excelsior*) sprout up a week earlier than specimens away from the light source. But urban heat island also contributes to this, namely, phenomenon that the average temperature in cities is a little higher. Lamp light also affects leaf movement, stomatal conductance, abaxial and adaxial cell elongation, cytoskeletal rearrangement, carbon dioxide uptake, hormone production and transport [52].

Another effect on plants is the reduction in the population size of pollinating insects, which is drastically decreased as a result of artificial light. Moths are important nocturnal pollinators, and due to the strong decline in their population, pollination fails often, so this indirect effect also threatens the healthy functioning of the ecosystem [53].

During the dark period, regeneration processes begin in plant cells to repair damages occur during the day. In this case, the normal physiological state is restored (Light-Independent Repair) [54]. Urban lighting makes it harder for plants to repair damage caused by oxidative stress. Ozone concentration in the air is also often high in cities due to pollution. Of course, ozone in high concentration also damages plant tissues, but the light pollution makes regeneration much more difficult for them [55]. Survival in a polluted urban environment is also made difficult for plants by abnormal lighting conditions (e.g. [56]).

Author details

Levente Hufnagel[1*] and Ferenc Mics[2]

1 Research Institute of Multidisciplinary Ecotheology, John Wesley Theological College, Budapest, Hungary

2 Department of Environmental Security, John Wesley Theological College, Budapest, Hungary

*Address all correspondence to: leventehufnagel@gmail.com

References

[1] Gimesi, L, Hufnagel, L. (2010): THE POSSIBILITIES OF BIODIVERSITY MONITORING BASED ON HUNGARIAN LIGHT TRAP NETWORKS - Applied Ecology and Environmental Research 8(3): 223-239.

[2] Nowinszky, L. (2004): NOCTURNAL ILLUMINATION AND NIGHT FLYING INSECTS- Applied Ecology and Environmental Research 2(1): 17-52

[3] Nowinszky, L, Puskás, J., Kúti, Zs. (2010): Light trapping as a dependent of Moonlight and clouds - Applied Ecology and Environmental Research 8(4): 301-312.

[4] Nowinszky, L, Puskás, J., Tar, K, Hufnagel, L, Ladányi, M. (2013): The dependence of normal and black light type trapping results upon the wingspan of moth species- Applied Ecology and Environmental Research 11(4): 593-610.

[5] EIA. (2018): Electric Sales, Revenue, and Average Price - Energy Information Administration. viewed on 28 january 2021. https://www.eia.gov/electricity/sales_revenue_p rice/.

[6] Tatro, K. (2020): Light Energy: Our Wasted Resource. – Consilience 22: 65-72.

[7] Verheijen, F.J. (1985): Photopollution: artificial light optic spatial control systems fail to cope with. Incidents, causations, remedies. – The Journal of Experimental Biology 44(1): 1-18.

[8] Li, X., Zhou, Y., Zhao, M. (2020): A harmonized global nighttime light dataset 1992-2018. – Scientific Data 7: 168.

[9] Bará, S., Lima, R.C., Zamorano, J. (2019): Monitoring Long-Term Trends in the Anthropogenic Night Sky Brightness. – Sustainability 11(11): 3070.

[10] Bennie, J., Duffy, J.P., Davies, T.W., Correa-Cano, M.E., Gaston, K.J. (2015): Global Trends in Exposure to Light Pollution in Natural Terrestrial Ecosystems. – Remote Sensing 7: 2715-2730.

[11] Brighenti, L.S., Staehr, P.A., Gagliardi, L.M., Brandão, L.P.M., Elias, E.C., de Mello, N.A.S.T., Barbosa, F.A.R., Bezerra-Neto, J.F. (2015): Seasonal Changes in Metabolic Rates of Two Tropical Lakes in the Atlantic Forest of Brazil. – Ecosystems 18: 589-604.

[12] Sachsenweger, M., Sachsenweger, U. (2012): Stereoscopic acuity in ocular pursuit of moving objects: Dynamic stereoscopy and movement parallax: relevance to road safety and occupational medicine. – Springer Science & Business Media, Dordrecht.

[13] Donner, K., Zak, P., Viljanen, M., Lindström, M., Feldman, T., Ostrovsky, M. (2016): Eye spectral sensitivity in fresh- and brackish-water populations of three glacial-relict *Mysis* species (Crustacea): physiology and genetics of differential tuning. – Journal of Comparative physiology A, Neuroethology, Sensory, Neural, and Behavioral Physiology 202(4): 297-312.

[14] Schweikert, L.E., Fitak, R.R., Caves, E.M., Sutton, T.T., Johnsen, S. (2018): Spectral sensitivity in ray-finned fishes: diversity, ecology and shared descent. – Journal of Experimental Biology 221: jeb189761.

[15] Owens, A.C.S., Lewis, S.M. (2018): The impact of artificial light at night on nocturnal insects: A review and synthesis. – Ecology and Evolution 8: 11337-11358.

[16] Gaston, K.J., Visser, M.E., Hölker, F. (2015): The biological impacts of artificial light at night: the research challenge. – Philosophical Transactions of the Royal Society B 370(1667): 20140133.

[17] Schroer, S., Hölker, F. (2016): Impact of Lighting on Flora and Fauna. – In: Karlicek, R., Sun, C.-C., Zissis, G., Ma, R. Handbook of Advanced Lighting Technology. Springer International Publishing, Switzerland.

[18] Leveau, L.M. (2020): Artificial Light at Night (ALAN) Is the Main Driver of Nocturnal Feral Pigeon (*Columba livia* f. *domestica*) Foraging in Urban Areas. – Animals (Basel). 10(4): 554.

[19] Zozaya, S.M., Alford, R.A., Schwarzkopf, L. (2015): Invasive house geckos are more willing to use artificial lights than are native geckos. – Austral Ecology 40(8): 982-987.

[20] Da Silva, A., Kempenaers, B. (2017): Singing from North to South: Latitudinal variation in timing of dawn singing under natural and artificial light conditions. – Journal of Animal Ecology 86(6): 1286-1297.

[21] Dimitriadis, C., Fournari-Konstantinidou, I., Sourbes, L., Koutsoubas, D., Mazaris, A.D. (2017): Reduction of sea turtle population recruitment caused by nightlight: Evidence from the Mediterranean region. – Ocean & Coastal Management 153: 108-115.

[22] Contín, M.A., Benedetto, M.M., Quinteros-Quintana, M.L., Guido, M.E. (2016): Light pollution: the possible consequences of excessive illumination on retina. – Eye 30(2): 255-263.

[23] Ronconi, R.A., Allard, K.A., Taylor, P.D. (2015): Bird interactions with offshore oil and gas platforms: Review of impacts and monitoring techniques. – Journal of Environmental Management 147: 34-45.

[24] Barroso, A., Haifig, I., Janei, V., da Silva, I., Dietrich, C., Costa-Leonardo, A.M. (2015): Effects of flickering light on the attraction of nocturnal insects. – Lighting Research & Technology 49(1): 1-11.

[25] Gomes, D.G.E. (2020): Orb-weaving spiders are fewer but larger and catch more prey in lit bridge panels from a natural artificial light experiment. – PeerJ 8:e8808.

[26] Okland, B. (2002): Canopy cover favours sporocarp-visiting beetles in spruce forest. – Norwegian Journal of Entomology 49: 29-39.

[27] Shimoda, M., Honda, K. (2013): Insect reactions to light and its applications to pest management. – Applied Entomology and Zoology 48(4).

[28] Elvidge, C.K., Ford, M.I., Pratt, T.C., Smokorowski, K.E., Sills, Patrick, P.H., Cooke, S.J. (2018): Behavioural guidance of yellow-stage American eel *Anguilla rostrata* with a light-emitting diode device. – Endangered Species Research 35: 159-168.

[29] Halfwerk, W., Blaas, M., Kramer, L., Hijner, N., Trillo, P.A., Bernal, X.E., Page, R.A., Goutte, S., Ryan, M.J., Ellers, J. (2019): Adaptive changes in sexual signalling in response to urbanization. – Nature Ecology & Evolution 3: 374-380.

[30] Raap, T. (2018): Effects of artificial light at night on the behaviour and physiology of free-living songbirds. – PhD thesis, University of Antwerpen, Belgium.

[31] Lewis, S.M., Wong, C.H., Owens, A.C.S., Fallon, C., Jepsen, S., Thancharoen, A., Wu, C., De Cock, R.,

Novák, M., López-Palafox, T., Khoo, V., Reed, J.M. (2020): A Global Perspective on Firefly Extinction Threats. – BioScience 70(2): 157-167.

[32] Birnie-Gauvin, K., Peiman, K.S., Gallagher, A.J., de Bruijn, R., Cooke, S.J. (2016): Sublethal consequences of urban life for wild vertebrates. – Environmental Reviews 24(4): 416-425.

[33] McMahon, T.A., Rohr, J.R., Bernal, X.E. (2017): Light and noise pollution interact to disrupt interspecific interactions. – Ecology 98(5): 1290-1299.

[34] Buchanan, B.W. (1998): Low-illumination prey detection by squirrel treefrogs. – Journal of Herpetology 32: 270-274.

[35] Hailman, J.P. (1984): Bimodal nocturnal activity of the western toad (*Bufo boreas*) in relation to ambient illumination. – Copeia 1984: 283-290.

[36] Plummer, K., Hale, J., O'Callaghan, M., Sadler, J., Siriwardeni, G. (2016): Investigating the impact of street lighting changes on garden moth communities. – Journal of Urban Ecology 2(1): juw004.

[37] Stone, E.L., Harris, S., Jones, G. (2015): Impacts of artificial lighting on bats: a review of challenges and solutions. – Mammalian Biology 80: 213-219.

[38] Bennie, J., Davies, T.W., Cruse, D., Inger, R., Gaston, K.J. (2015): Cascading effects of artificial light at night: resource-mediated control of herbivores in a grassland ecosystem. – Philosophical Transactions of the Royal Society B: Biological Sciences 370: 20140131.

[39] Farnworth, B., Meitern, R., Innes, J., Waas, J.R. (2019): Increasing predation risk with light reduces speed, exploration and visit duration of invasive ship rats (*Rattus rattus*). – Scientific Reports 9: 3739.

[40] Berge, J., Geoffroy, M., Daase, M., Cottier, F., Priou, P., Cohen, J.H., Johnsen, G., McKee, D., Kostakis, I., Renaud, P.E., Vogedes, D., Anderson, Last, K.S., Gauthier, S. (2020): Artificial light during the polar night disrupts Arctic fish and zooplankton behaviour down to 200 m depth. – Communications Biology 3: 102.

[41] Yurk, H., Trites, A.W. (2000): Experimental attempts to reduce predation by harbor seals on out-migrating juvenile salmonids. – Transactions of the American Fisheries Society 129: 1360-1366.

[42] Yorzinski, J.L., Chisholm, S., Byerley, S.D., Coy, J.R., Aziz, A., Wolf, J.A., Gnerlich, A.C. (2015): Artificial light pollution increases nocturnal vigilance in peahens. – PeerJ 3:e1174.

[43] Wakefield, A., Stone, E.L., Jones, G., Harris, S. (2015): Light-emitting diode street lights reduce last-ditch evasive manoeuvres by moths to bat echolocation calls. – Royal Society Open Science 2: 150291.

[44] Last, K.S., Hobbs, Berge, J., Brierley, A.S., Cottier, F. (2016): Moonlight Drives Ocean-Scale Mass Vertical Migration of Zooplankton during the Arctic Winter. – Current Biology 26: 1-8.

[45] Luarte, T., Bonta, C.C., Silva-Rodriguez, E.A., Quijon, P.A. Miranda, C., Farias, A.A., Duarte, C. (2016): Light pollution reduces activity, food consumption and growth rates in a sandy beach invertebrate. – Environmental Pollution 218: 1147-1153.

[46] Mathews, F., Roche, N., Aughney, T., Jones, N., Day, J., Baker, J., Langton, S. (2015): Barriers and benefits: implications of artificial night-lighting for the distribution of common bats in Britain and Ireland. – Philosophical

Transactions of the Royal Society B: Biological Sciences 370: 20140124.

[47] Andreani, T.S., Itoh, T.Q., Yildirim, E., Hwangbo, D.S., Allada, R. (2015): Genetics of circadian rhythms. – Sleep Medicine Clinics 10(4): 413-421.

[48] Sreenivasulu, N. (2017): Systems biology of seeds: deciphering the molecular mechanisms of seed storage, dormancy and onset of germination. – Plant Cell Reports 36(5): 633-635.

[49] van Gelderen, K., Kang, C., Pierik, R. (2017): Light Signaling, Root Development, and Plasticity. – Plant Physiology 176: 1049-1060.

[50] Škvareninová, J., Tuhárska, M., Škvarenina, J., Babálová, D., Slobodníková, L., Slobodník, B., Středová, H., Minďaš, J. (2017): Effects of light pollution on tree phenology in the urban environment. – Moravian Geographical Reports 25(4): 282-290.

[51] French-Constant, R.H., Somers-Yeates, R., Bennie, J., Economou, T., Hodgson, D., Spalding, A., McGregor, P.K. (2016): Light pollution is associated with earlier tree budburst across the United Kingdom. – Proceedings of the Royal Society B: Biological Sciences 283(1833): 20160813.

[52] Roni, Z.K., Islam, S., Shimasaki , K. (2017): Response of *Eustoma* Leaf Phenotype and Photosynthetic Performance to LED Light Quality. – Horticulturae 3(4): 50.

[53] MacGregor, C.J., Pocock, M.J.O., Fox, R., Evans, D.M. (2015): Pollination by nocturnal Lepidoptera, and the effects of light pollution: a review. – Ecological Entomology 40(3): 187-198.

[54] Gill, S.S., Anjum, N.A., Gill, R., Jha, M., Tuteja, N. (2015): DNA Damage and Repair in Plants under Ultraviolet and Ionizing Radiations. – The Scientific World Journal Article ID 250158.

[55] Meravi, N., Prajapati, S.K. (2018): Effect street light pollution on the photosynthetic efficiency of different plants. – Biological Rhythm Research 51(1): 67-75.

[56] Yang, J, Duan, R. (2019): The effect of artificial illumination on postponing plant phenology – Applied Ecology and Environmental Research 17(6):13289-13296

Chapter 2

Artificial Light at Night: A Global Threat to Plant Biological Rhythms and Eco-Physiological Processes

Rekha Sodani, Udit Nandan Mishra, Subhash Chand, Indu, Hirdayesh Anuragi, Kailash Chandra, Jyoti Chauhan, Bandana Bose, Vivek Kumar, Gopal Shankar Singh, Devidutta Lenka and Rajesh Kumar Singhal

Abstract

Light is crucial environmental factor for primary resource and signalling in plants and provide optimum fitness under fluctuating environments from millions of year. However, due to urbanization, and human development activities lot of excess light generated in environment during night time and responsible for anthropogenic generated pollution (ALAN; artificial night light pollution). This pollution has cause for serious problem in plants as it affects their processes and functions which are under the control of light or diurnal cycle. Plant biorhythms mostly diurnal rhythms such as stomatal movements, photosynthetic activity, and many more metabolic processes are under the control of period of light and dark, which are crucially affected by artificial light at night. Similarly, the crucial plant processes such as pollination, flowering, and yield determining processes are controlled by the diurnal cycle and ALAN affects these processes and ultimately hampers the plant fitness and development. To keep in mind the effect of artificial light at night on plant biorhythm and eco-physiological processes, this chapter will focus on the status of global artificial night light pollution and the responsible factors. Further, we will explore the details mechanisms of plant biorhythm and eco-physiological processes under artificial light at night and how this mechanism can be a global threat. Then at the end we will focus on the ANLP reducing strategies such as new light policy, advanced lightening technology such as remote sensing and lightening utilisation optimisation.

Keywords: artificial night light pollution, minimizing strategies, photosynthesis, pollination, plant growth and development, plant movements

1. Introduction

Light is a major abiotic factor acts as energy source and signalling for plants growth and developments. Sun light is the prime source of energy on this planet and regulates number of essential functions in living organisms. Among them, primary producer as plants, blue green algae and photoautotrophs absorb these lights (especially 400-700 nm refereed as photosynthetically active radiation) and convert into energy source (starch, sucrose and other complex organic compound as food reserve) through the mechanism of photosynthesis [1]. Moreover, other living organisms like heterotrophs (primary consumer and secondary consumers) fulfil their energy requirements via feeding the primary producer food reserves. Therefore, light is an essential factor for all living organism to fulfil their energy requirements. Another important function of light is to act as regulator of signalling along with growth and development functions of plant. For example, photomorphogenesis (light regulated morphogenesis), phototropism (light mediated movements), circadian rhythm (light regulated biological rhythms) and many other crucial processes are under control of light [2]. Light characteristics such as intensity, duration, and wavelength affect the living organisms including plant in positive and negative way. Similarly, from millions of years plants are adaptive themselves to the diurnal changes of light–dark timings and many of the research concluded that change in light–dark duration affect plant growth mild to very drastic levels [3]. However, recent advancement in human development such as highway, buildings, LEDs lightening technology, product advertisements and industries progression lead to generation of excess light in the surrounding which causes anthropogenic accelerated light pollutions. Unlike natural ecosystems, where daily activities are scheduled by natural light–dark diurnal cycles [4], cities are heavily lit to enable performance of a wide array of activities after dark [5]. As artificial light at night (ALAN) becomes more reliable, efficient and affordable, living organisms become increasingly exposed to drastic and pervasive effects of "light pollution."

According to Encyclopaedia Britannica, "light pollution" is "unwanted or excessive artificial light" during night hours. The introduction of ALAN into the environment majorly through electric light sources related to domestic purpose, industrial areas, transportation and street lights. The artificial night light altered the natural light cycle through its beneficial along with harmful effects on the plant ecosystem. Its beneficial impacts imply reduced the risk of night-road accident, the crime rate at night (approximately 30%), and increased working hours and scientific research areas such as speed breeding and tissue culture [6]. Therefore, there is a need to balance the positive and negative impacts of ALAN. Biology can be a science of timing/duration, for example all organisms experience the drastic results due to change in natural light cycle variations [6]. They are the utmost important physical factors for time. These include common daily activities (e.g. photosynthetic activity, stomatal movements, enzyme activity, flower opening, sleep movements, fragrance emission, dark recovery and repair). Indeed, the lunar cycle and the yearly/seasonal cycle dominate and regulate the lives of most organisms. Influences on biological life history have repeatedly been found to create events that occur later in life. Climatic changes can cause food and reproductive capacity to become mismatched, which in turn leads to organisms projecting highly developed phenology [7]. Likewise, plant-environment interaction is very important to decide the present and future growth of living organisms including plants. It has been documented that ALAN affects the micro-environment attributes such as light, soil biology, humidity, biotic community and their interactions surrounding the living organisms including plant [8, 9]. Moreover, in few recent studies it was showed that ALAN not only limited to light area but it can affect the living organism beyond this and can drastically affects the ecosystem services and biodiversity [10]. Consequently, pollination, net primary productivity, flowering, ecosystem services and nutrient recycling are the important eco-physiological functions, which are influenced by ALAN. The effect of ALAN in living organisms such as animals and humans are studied very well and several studies, and meta-analyses showed the effects on as behaviour (sleep, food, foraging, and flying), reproduction, vigilance and many other important activities [11, 12]. However, the effect of ALAN are very limited till date, only in last few years scientists are working on the ALAN and plant processes relationships and achieved some milestones. While considering the above facts, this chapter covers the present advances in ALAN research on plants specific to biorhythms and eco-physiological functions. Consequently, this analysis point out on some important strategies which are crucial for minimizing the extent of this pollution in environment.

2. Artificial light at night (ALAN) current status and anthropogenic sources of ALAN

Human development is continuous and complex process, also crucial for their existence for long term on this planet. This process culminates a number of novel pollutants (also known as anthropogenic pollutants) and creates a problem of other living organisms also environment sustainability. Although, this emerging pollution is global problem but the conditions might be severe when the pollution affects the endangered species in protected areas and natural hotspots [13]. Therefore, researchers tried to quantifying the ALAN worldwide by data generated from remote sensing technology, geographical information system, hyperspectral, visual infrared imaging radiometer, day/night band, and satellites [14–16]. In the year 2001, the first globe atlas of artificial night sky brightness based on US air force satellite data, which reported that the brightness of sky is increasing continuously and in world nearly two-third of world population living in an area where ALAN is higher than threshold limits (when the artificial night light brightness is greater than 10% of natural brightness). Also, many countries such as US had this value much higher and above 99% population facing this problem [17]. Consequently, the new world atlas on artificial night light brightness by using satellite data, day/night band, and Suomi national polar orbiting partnership satellite, which improve the resolution and accuracy reported that in new world 80% of world population facing the problem of excess night light brightness (brightness >14μcd/m^2) [15]. Therefore, ALAN is spreading swiftly and considered as global problem and now every countries trying to study the trend of light brightness in their protected area using advanced technology. For example, India analyse the trend from 1993–2013 using defense meteorological satellite programme, which help in identifying the sensitive sites such as protected areas and setting the new light policies and priorities [18]. Although the ALAN varies in intensity, colour, timing, and wavelengths and also vary with the particular country and location. A list of ten highly polluted night light brightness cities of world and India are showed in **Figure 1(a, b)**. In the represented figures, data on light brightness using light pollution map reveals that the situation of some cities in world such as Doha, Abu Dhabi, and Kuwait are very crucial as they have light brightness >160nW/cm^{2}*Sr., which is 3–4 times higher than highly night light polluted brightness (50nW/cm^{2}*Sr. considered as highly light polluted area).

The natural sky glow during night consists of comes from moon light, integrated starlight's, zodiacal light and airglow. In the recent times due to anthropogenic activities excess light generated in the surrounding that causes disappearance of natural darkness, artificial glow in sky, and loss in visibility of the stars and Milky Way's [19]. There are several anthropogenic factors that are responsible for pollution in night light such as decorative lightening during various festivals, lightening in buildings, malls and in homes, traffic light (consist of four wheeler, two wheeler light), road/highway light (to reduce accidents), advertisement light (to enhance marketing of different products), ship and aeroplane light, and street

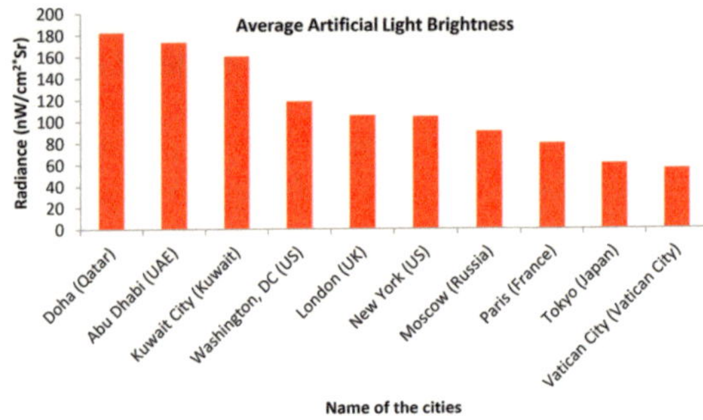

a

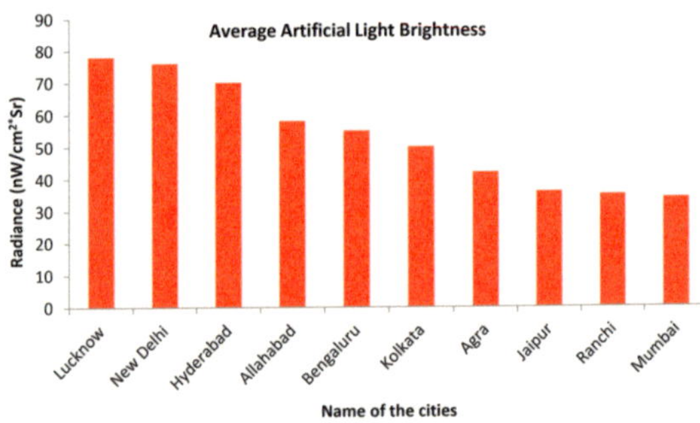

b

Figure 1.
(a, b) Represent the list of top ten world highly night light brighten cities (mean value of 28.27 km^2 area at 10 km elevation) of world, (highest value is near to 185 which is very high), and India during the year 2019-2020. Data procured from https://www.lightpollutionmap.info/.

lightening [20]. These artificial lights vary in wavelength, timing (few minutes to whole night), colour (violet to red) and intensity (very low to very high). Also, the origin of this lightening varies such as LEDs light (light emitting diode), halogen light, tungsten bulb, and florescent light (tube lights). Moreover, it varies with the locations such as market area having light for 3 to 4 hours, home lights early evening hours and road light (whole night).

3. Plant biorhythms or circadian rhythms and their response under light and dark cycle

Biorhythms are known as adjustment mechanism in living systems that keep body homeostatis, adaptive processes and dynamic equilibrium. Biological rhythms are endogenous and closely linked with external environmental elements but regulated by endogenously such as genes. Phytochrome and cryptochrome pigment plays a key role in coordination and regulation of plant biorhythms [21, 22]. Phytocrome repons to red and far-red light, while crytochrome responds to ultra violet/blue light. Arabidopsis has five phytochrome *PHYA* to *PHYE* and two crytochrome *CRY1* and *CRY2* [21, 22]. Plant biorhthms are mostly classified as (a) cicadian rhythms; (around or approximately; dies meaning day), (b) circaseptan rhythms; (weekly), (c) circalunar rhythms; (monthly), (d) circannual rhythm; (annual) [21, 23, 24]. In different way, biorhythms classified on the basis of duration for example, (a) short duration; on the order of seconds or minutes for example water and ion exchange across the cell membranes, (b) long duration; varies from days to months for example complex sequence of events and the quantitative changes regulated at cell, tissue and organ level [25–28]. Moreover other classification of biorthyms include (a) ecological rhythms; monthly, seasonal or diurnal (b) Physiological rhythms; based on various organs bioelectric activity [29–30].

Among the all known rhythms, circadian thythms are very well known and documented by many reserchers. Circadian clocks are the biological oscillators that enable the organism to coordinate their physiology and behavior under periodic environmental fluctuations and also evolved in organisms in response to the daily rotation of the earth [23]. Circadian clock can influence diverse plant crucial processes like leaf movement, photosynthesis, stem extension, stomatal opening and hormonal regulations [23]. The circadian clock constitutes of three components: input, central oscillator and output pathways. Each component contains a number of genes i. e., PHYs (PHYTOCHROMEs), CRYs (CRYPTOCHROMEs) and PHOTs (PHOTOTROPINs) are the very well recognized light receptor genes of input pathways and transmit external light stimuli into the central oscillator [31]. Phytochrome interacting factor 3 (*PIF3*), ZEITLUPE (ZTL), PSUEDO-RESPONSE REGULATOR (*PRR*) have important role during light signalling and affects the circadian clock [32]. Free running period (FRP) is the period length of cercadian rhythms measured under constant conditions and varies among tissues, organisms and even cells of the same cell type [32]. In circadian of *Arabidopsis thaliana* oscillator consist of interlocking transcriptional feedback loops and control significant process such as growth and metabolism. Circadian rhythms defined by three fundamental parameters: Periodicity; time to complete one cycle of 24 hours [33], Entrainability; circadian rhythms are self-sustaining and endogenously generated, therefore they maintain under constant environmental conditions like constant light or dark and temperature, Temperature compensation; the period remain relatively constant over a range of ambient temperature.

The circadian rhythm is closely associated to the light–dark cycle. Circadian rhythms remain consistent in response to no time cues but can be entrained by ambient conditions. But it's been clear that different stimuli have varying effects on our circadian rhythms. Many environmental parameters provide stimulus to the clock, where the best characterized and most potent entraining stimulus is light in plants [34, 35]. Further, it is shows that shoot tip sends an unknown signal to the roots so root maintain circadian rhythm. While the lack of shoot apex signals loss of rhythmicity in the decapitated root, which can be overcome by direct exposure of root to LD cycles (light: dark) even as low intensities. It is clear that roots are entrained by light in preference to shoot apex derived signals [36]. In the context of photoperiodism, the circadian rhythm is combined with light signaling. The photoperiod sensor permits plants to respond to the annual cycle of day length, by the production of flowers, tubers and frost tolerant buds in appropriate seasons [37]. Accurate entrainment is important for photoperiodism, certainly general physiology shows that the crucial difference between light dominant plants (most of the flowering in long days) and dark dominant plants (most flowering in short days) is in the entrainment of their photoperiodic rhythm [38]. In light signaling pathways both phytochrome (phy) and cryptochrome (cry) regulate clock components to attain entrainment in plants [39]. The phytochrome responds to a red light and the cryptochrome absorbs in the UV-A/blue wavelength. Recognition of variations in day length confers plants seasonal flowering. This mechanism includes a time-keeping mechanism that integrates intimation of light environment to estimate the duration of day or night. Time-keeping activity is the outcome of the circadian clock. In Arabidopsis, an increase in flowering under long days (LD) happens through transcriptional induction of florigen gene FLOWERING LOCUS T (FT) specifically under LD conditions. The FT promoter binds with CONSTANS (CO) transcription factor which directly confer its LD-specific induction [40].

CONSTANS characterizes as a photoperiodic timer gene with its transcript level being managed by the circadian clock. CO proteins accumulate in response to exposure to light [41].

4. Impact of the artificial light at night time on circadian rhythms

As sessile plant receive light as signal and resource. Diurnal cycle of dark and light period is vital for regulating numerous processes in plants such as daily events of enzymes activity, gas exchange, photoperiodic movements, metabolism, stomatal movements, flowering opening, dark repair recovery and photosynthesis. The impact of artificial lighting on the body clock and its circadian rhythms is an important research topic. Night shifts introduction has changed the natural 24-hour cycle. There are multiple different categories of artificial lighting that are being utilized. It is various forms, from brief pulse to long lasting night glow, from narrow emission spectra to broad emission spectra, from low emission intensity to high intensity emission, and from local emission focus to glowing of sky. Effect of ALAN on some plant circadian process are disused thoroughly and represented in **Figure 2**.

4.1 Enzyme activity or metabolism

It is very well documented that enzymes are essential for optimum growth and development of plant. Also, they help in regulating all the physiological, biochemical processes in plants. Important physiological and biochemical processes of plant include such as seed germination, photosynthesis, respiration, and translocation. In this regards an experiment conducted in rice using the different light intensity (from low to high) during night hours and observed that ALAN drastically reduce the germination capacity by reducing the activity of α-amylase enzyme [42]. Moreover, a recent study conducted perennial ryegrass using the different duration

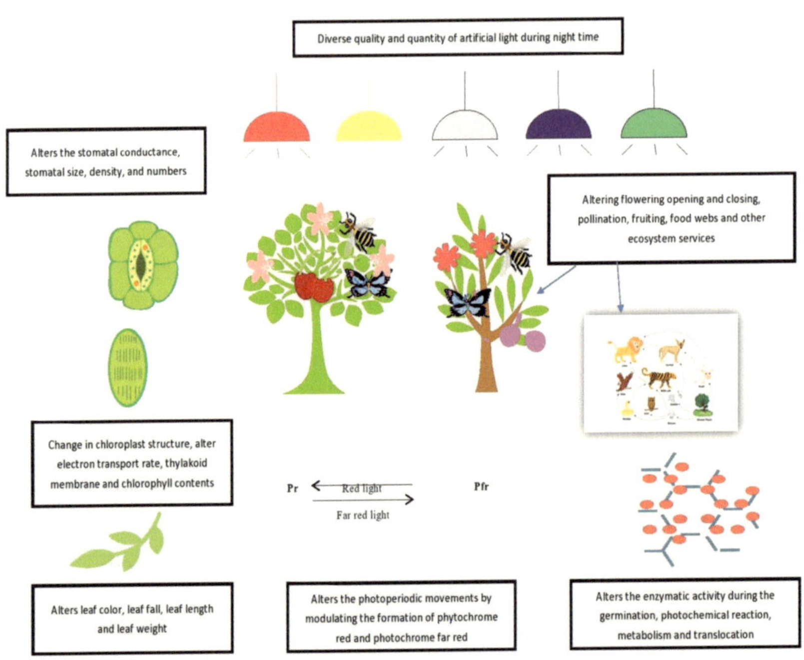

Figure 2.
Represents the impact of artificial light at night on plant circadian and eco-physiological processes.

of light as light/dark hr. (24/0, 22/2, 20/4, 18/6, 16/8, and 14/10) and they had found that seed germination percentage (67 to 33%), soluble carbohydrates (27.48 to 9.16 mg/g fresh weight) and soluble protein (13.85 to 10.59 mg/g fresh weight) are lowest in 24/0 conditions as compared to 16/0 light/dark conditions, which showed the drastic effect on the future growth of plant [43]. In this array, a study conducted in yellow poplar using different light intensity from low 1 to high 50 $\mu mol\ m^{-2}\ s^{-1}$ showed that in ALAN leading to reduction in starch turnover (74.7 to 11.4%) via affecting the starch synthesis activity and lower rate of respiration [44, 45]. Therefore, from these studies, it can be concluded that ALAN affect the metabolism of plant at early and late growth stages.

4.2 Stomatal movement and biomass

The stomata are very important for gas exchange and water transpiration in plants. It is also well documented that stomatal movement is circadian rhythm which is also influenced by the intensity and quality of light. In this regard, an experiment conducted using different photon flux density (1, 3, and 50 $\mu mol\ m^{-2}\ s^{-1}$) and different light sources low pressure sodium lamp, high pressure sodium lamp and LEDs and they found shorter, narrower stomatal aperture, change in osmotic pressure and reduced stomatal size with increased stomatal density and the higher intensity more drastically affect the stomatal attributes in comparison with lower intensity of ALAN. Further, it is also found that the stomatal conductance is reduced under ALAN condition and affects the biomass accumulation and gas exchange [44, 45]. ALAN reduces the both above and below ground fresh and dry weight. Such as in the presence of 50 $\mu mol\ m^{-2}\ s^{-1}$ in ryegrass during night time reduces the below ground fresh and dry weight by 48 and 46% respectively. Similarly, the reduction in above ground fresh and dry weight are 27 and 46% respectively. Therefore, it is expected that in future the result of ALAN might be more drastic and influence stomatal behaviours in plants.

4.3 Photosynthesis

Photosynthesis is the most crucial physiological process in autotrophs and essential for sustaining life on this planet. Earlier, it is documented that the sunlight is the prime source of energy for photosynthesis processes but in recent years many studies conducted using ALAN and found that it hampers the photosynthetic processes. In this aspect a recent experiment conducted using the street light having ALAN intensity near to 340-360lux and exhibited that plant under street light have lower photosynthetic quantum yield (Fv/Fm), non-photochemical quenching (NPQ) values, which ultimately affects the photosynthetic rate drastically [46]. Similarly, a study conducted in yellow poplar using different light intensity and quality during night hours reveals that total chlorophyll content reduced up to 35%, water use efficiency 23%, and photosynthetic rate 42 to 45% [45]. Another study reported in yellow poplar showed that the ALAN affect the ultrastructure of chloroplast by multiplying number of thylakoid membrane stack, reducing thylakoid stacking, and increasing number and size of plasoglobuli [44]. These modifications in chloroplast ultrastructure lead to early senescence, chloric and abnormal leaves. Consequently, ALAN in ryegrass influences the Chl a, chl b and total chlorophyll content, Fv/Fm ration, and electron transport chain. Further, they also ended up that the ALAN act as stressor and influence the photosynthetic efficiency of plants [43]. Therefore, it can be concluded that ALAN act as a stressor for photosynthetic phenomena and expected that in future it may be a global problem of road side trees.

4.4 Photoperiodic movements

The impact of relative length of flowering during day and night is refereed as photoperiodism. On the basis of day and night length plants are categorised into long day plant (required longer day length than critical period), short day plant (required shorter than crucial photoperiod) and day neutral plant (do not affect by day length). Therefore, flowering mechanism is very sensitive to critical day length. In the recent years it is found that ALAN severely affects the flowering mechanism in plants. For example, under ALAN and average spring temperature conditions at timing of budburst in deciduous tree species recorded and observed the budburst occurrence about 7.5 days earlier as compared to the normal UK conditions [47]. A long term experiment conducted to observe the impact of ALAN on wild species in natural and semi-natural grassland and authors found that the lightening affects the trajectory of vegetation cover, leading to change in plant biomass and the composition of dominant wild species. Further, they also suggested that the ALAN significantly alters the flowering phenology by shortening or lengthening of flowering period varies from 4 day earlier to 12 day late compared to control conditions [48]. Nevertheless, authors observed the non-significant relationship of ALAN with vegetation composition and flower density in grassland vegetation species. Further, they observed early flowering (about 4 days) in *Agrostis tenuis* and delayed flowering (about 7 days) in *Anthoxanthum odoratum* [49]. Thus, flowering behaviour of the plant species varies with ALAN (timing, duration, intensity, and spectrum etc.). Similarly, a study conducted on foredune vegetation such as *Traganum moquinii* receiving various light intensities during the year and author suggested that vegetation which face direct light or close to light source affected more and their flower and seed production is reduced drastically compared to species which located far from light source [50]. Also the flowering period can vary (reduced or increased) under ALAN, depend upon the type of plant species with their genetic constituents. For example, an experiment conducted using 1000-4000 lx light using LEDs and found that trees flowering are more susceptible than shrub species and plant species are more susceptible than evergreen species [51]. Therefore, it might be possible that in future the photoperiodism phenomenon affected more and can be responsible for evolutionary changes. The responses of plant circadian and ecological processes are represented in **Figure 2**.

5. Impact of ALAN on eco-physiological processes

As earlier mentioned ALAN is a global problem and it act as a stressor for living organisms including plant. Also, some recent studies exhibited that the impact of ALAN can be amplified with unfavourable conditions as pollution. Moreover, the living organisms or plants connected with environmental phenomenon and adaptive themselves. Many plant processes are associated to environment such as transpiration which depends upon the atmosphere humidity. Some of eco-physiological processes influenced by ALAN condition are followed and represented in **Figure 2**.

5.1 Plant growth and development

It is very well documented that the growth and development of plant is highly influenced by the environmental conditions. Also, a positive plant-environment interaction is necessary to grow plant under different environmental circumstances. The night hour lightening has great impact on the plant phenology, physiology, growth and development, reproduction and behaviours [52]. The light pollution is an emerging global phenomenon and accelerated by rapid urbanization, which affects both plant and animal fitness in both developed and developing countries, however the impact are very serious in urban areas. Contrary, it is observed that delayed autumn phenophases in the crown part of two tree species (*Acer pseudo-platanus* L. and *Rhus typhina* L.) positioned next to light source by 13 to 22 days and also deferred duration by 6–9 days. They also observed delayed leaf colouring duration by 6 to 9 days and leaf fall by 6 to 7 days due to ALAN [53]. A recent study conducted using ALAN in perennial wildflower regarding to observe the effect on growth and developments and they concluded that ALAN modulates the growth by interaction with abiotic factors such as soil moisture and affect plant density [8]. Likewise, recent study on the ecophysiological responses of ALAN in plants concluded that this had potential to alter the number of aspects such as germination, photosynthesis, biomass and yield [54]. In these consequences a recent study conducted using the different light intensities during night in rice and authors found out that the harmful impacts of light during night increased with light intensities and drastically reduces the seed germination (−14%), seedling vigour (16.83 to 12.51), root length (8.63 to 7.03) and early growth attributes such as seedling length (17.72 to 13.72 cm), mean time of germination (1.21 to 1.28), coefficient velocity of germination (82.61 to 78.07), germination index (526 to 465), germination rate index (87.83 to 78.50%/day) and mean germination rate (0.83 to 0.78) under artificial light conditions as compared to control, which directly associated the future growth of plant [42]. Moreover, in ryegrass reported that different intensity of light affects the leaf length (20.82 to 16.19 cm), leaf weight (1.02 to 0.58 g/10plants) and physiological parameters, which ultimately affects the growth in negative ways [43]. Therefore, it is expected that ALAN significantly hampers the growth and development of plant.

5.2 Pollination

Globally, pollinators are indispensable source of pollination in both agriculturally important crops and wild plants [55]. Anthropogenic activities such as intensive agriculture, use of extensive pesticides, habitat change, invasive alien species and climate change has substantially reduced population of pollinators [56]. Moreover, ALAN has emerged a new threat to plant reproductive success by disturbing plant-pollinator ecosystem balance. For example, authors witnessed the reduction of nocturnal pollinators visit by 62% in ALAN areas compared to the dark areas and yield was reduced by 13% [57]. They also observed visit of diurnal pollinators at night with nocturnal pollinators under ALAN. The pollinator-population dynamics and plant-pollinator communities are being adversely affected by artificial night lightening. Furthermore, ecological functionality and stability has been challenged by artificial night lightening throughout the world. Also the pollution effect of ALAN by nocturnal moths (Lepidoptera) and concluded that pollination is an ecosystem driven process that can be disturbed by increase of light pollution [58]. Similarly, recent study observed the negative relationship between direct ALAN and reproductive out in *Epilobium angustifolium*, plant species. This result can be due to disturb of pollination services in plants at nigh time by direct light illumination and cannot be compensated by day time pollinators [59]. Antagonistically, it is reported that there is no direct effect of light pollution on the reproductive output of *Silene latifolia* and possible only if diurnal pollinators compensate the decrease in pollination at night and another reason can be development of mechanism in pollinators which can manage the adverse effect of light pollution [59]. Thus, effect and extent of ALAN varies with plant species, and the behaviour of pollinators. Further, it can affect plant and animal biodiversity, species interaction particularly plant and pollinators, ecosystem balance and functioning.

5.3 ALAN as stressor

As sessile in nature plant faces a number of stressor throughout their life cycle such as abiotic stress (drought, heat, flood, cold, salinity, elevated CO_2, and heavy metals), pollutions (air pollution, soil pollution), biotic stresses (insect, and pathogen), and most of times these stress are in combined from and cause for detrimental loss in their optimum growth and developments. All of these are act as stressor for plant and affect plant by changing the physiological, biochemical and molecular processes [60]. Likewise, this light also an important environmental factor and low and high light both act as stressor [42, 61]. Recent studies shows that ALAN also act as stressor and found out that it increases the lipid peroxidation, and reduces total antioxidant capacity in autotrophic red sea corals [62]. Further, an experiment in perennial ryegrass showed that ALAN increases the malondialdehyde (MDA) content and reduce leaf heat stability [43]. Similarly, it is observed that yellow poplar plant accumulate hydrogen peroxide, superoxide radicles and reduced abscisic content under ALAN conditions [44, 45]. Therefore, from these studies it may be concluded that ALAN act as a stressor and responsible for oxidative stress.

5.4 Ecosystem services under ALAN

Abiotic and biotic component of ecosystem coordinate to each other which provide important ecosystem services to humans. Also, the stability of ecosystems is key for sustaining of life on this planet. However, in the recent past due to uncontrolled anthropogenic activities ecosystem and biodiversity services of ecosystems are losing continuously. Among the anthropogenic activities, ALAN is one of the swiftly expanding activity, which is now become a global problem for ecosystem and biodiversity services [10, 63]. It is suggested that ALAN affect the organisms flux across the ecosystems and key driver of ecosystem community structure and can modify the ecosystem functioning beyond the affected area [64]. Moreover, the effects of ALAN are not limited to plants but also it affects the aquatic, forest, desert, terrestrial, mountain and agriculture ecosystem from lower to higher levels [10, 59]. Further it has been found the ALAN had drastic impact of ecosystem services such as foraging, vision, reproduction, signalling and behaviour. Therefore, it can be concluded that ALAN had serious impact on ecosystem services of stable ecosystems.

6. Minimizing strategy of ALAN

6.1 Street light pollution

Street lights should be replaced with energy-efficient LEDs and proper shielding of light in streets and by managing their angels. Street lights should be installed where needed Indian standards [65] has classified the roads according to the traffic density. By considering the road category according to traffic density engineers can design the street light plan and provide the installation specifications. To a report from [66], the Bruhat Bengaluru Mahanagara Palike (BBMP) projected a plan to replace the existing 4.8 lakh city street lights with high energy-efficient LED lights. A number studies conducted to showing their impact on plants ecosystems and hot spots and most of cases it affect in negative way.

Although, some of researcher are concerned and documented about the impact of ALAN but there is no progress in changing the street lights. However, in recent time some countries follow the new light policies, where they replaced the old lights with more efficient lights and change the angle and height of light source to reduce the tree passes.

6.2 Buildings lights

This can be minimize using smart building architecture to decrease the use of excessive outdoor and indoor lights. The utilisation of LEDs, compact fluorescents (CFS) and warm coloured bulbs should be used to minimise energy use and by this way somehow we can protect our environment. The dimmers, motion-sensors and timers contribute a lot to energy saving. Turning-off the unnecessary indoor along with outdoor lights in houses and offices. Avoid Blue light during night time. International Dark Sky-Association (IDA) recommends use only warm lights for outdoor that includes low-pressure sodium (LPS), high-pressured Sodium (HPS) and low-colour LEDs. By using warm or filtered LEDs (CCT < 3,000 K; S/P ratio < 1.2) can reduce the blue light emission [67].

6.3 Government policies for ALAN

Artificial light pollution is now emerging pollution and governments has also considered it and have provided certain guidelines to decrease the energy wastage, damage to agro-ecological and wildlife ecosystem. During the 13th meeting of the Conference of the Parties, Gandhinagar, India Agenda No. 26.4 "the guidelines for light pollution for wildlife including marine turtle, seabirds and migratory shorebirds" under the Convention on the migratory species (prepared by the Government of Australia) [68]. By seeing the threat of sky glow and emerging challenges of increasing artificial light pollution to the marine species government has taken a proactive approach to develop artificial light in night pollution guidelines. These kinds of initiatives need to be taken by every country to combat artificial night light pollution. Artificial light at night became an unavoidable technology from the societies however this indispensable tool has harmful side effect in term of light pollution [69]. One can understand that artificial light at night is necessary to highlight the scenic beauties of the cities and to provide safety on roads however where artificial light at night is easily available at very low cost there it is over utilized which leads to discomfort in society with a view to disturbing the natural habitats, underline the beauty of skyline etc. Therefore now we can understand that ALAN has two phases of its representation. The detrimental effect can be seen through alteration in the sky glow which is an important constituent of biosphere [70]. According to one study, around one fifth of the global population affected by artificial sky brightness [17]. Challenge that we have is how we can use the ALAN in such a way that maximize the social benefits and reduce the impact of its pollution.

In this regards several articles are available at public domain [71–79]. They have summarized the solution like protecting species rich areas in their natural habitat, policy making on threshold and upper limit of light emission, environmental specific custom based light brightness which is adapted to vicinity of that area, time control of light emission, diurnal adapted colour spectrum and reducing the trespass of lightening. Apart from this, it is suggested that the five ways to minimize the harmful effect of ALAN *viz.* improving light fixtures to direct the light where it needed, switch the light off whenever it is not needed, dim the light and choose the appropriate Illuminants, colour spectra and filters, shade the light to protect the neighbor and learn from nature to maintain its sustainability [69]. Further, suggested to use the decreased height of light poles and increased spacing between light poles to reduce the pollution from ALAN [80]. It is also strongly recommended that environmental impact need to be considered in addition to energy consumption [81]. Moreover, considering the nature and landscape protection, protection of breeding sites and resting places of highly sensitive biological organism, assessing or measurement technology of ALAN pollution must improve to reduce the impact [82]. In addition to above explained policies author suggest that existing lighting plans must be evaluated and can be improved, reducing the decorative lightening, using of covered bulbs that face downwards, use of automatic system to turn off street light at certain times, outdoor lights with glare should be replaced with low glare, use of IDA approved light fixtures and use of motion sensors on important outdoor lights to reduce the impact of pollution caused by ALAN. Although, there are no specific light policy for reducing impact of street light on plant and living organisms but it may expected that in near future countries will adopt some new light policies. However, some of points can be considered under future light policies to reduce the impact of ALAN on plants, which are followed.

- Reduce the height of light source to reduce tress passes
- Change the angle and position of light source
- Light source could be away from trees
- Growing light insensitive tress and crops o road side areas such as day neutral species
- Avoid short day plants near to highways
- Promote self-pollinated species as compared to cross pollinated
- Avoid cross-pollinated species especially which nocturnal pollinated

This approach can be milestone to reduce the impact of ALAN on plants.

7. Conclusion

Human population is continuously expanding and degradation of natural resources is also increasing with human development activities. ALAN has important role during the human development process as it increases the working hours, work efficiency, reduce crime and accidents but due to excess use and accumulation in environment cause a problem of artificial night light pollution. ALAN pollution is anthropogenic and increasing swiftly and has global impact. In the presented chapter we tried to explore the impact of ALAN on plant biological rhythms and ecophysiological processes. Therefore, this study comprises of the recent status of ALAN in world and their sources. In this we concluded that many countries are facing the problem of ALAN and continue expanding their area. Further, we discussed about the plant biorhythms and their regulation by diurnal cycles and concluded that the plant biorhythms are highly sensitive too change in light/dark periods. Thereafter, we provide the some details regarding the plant circadian rhythm, which are affected by ALAN and concluded that ALAN had negatively associated with them. Later on, we detail the some ecophysiological functions under ALAN and concluded that ALAN had drastic effects on plant processes and in future it can be a global problem. In the last we discussed the some strategies and approaches to minimize the effect of ALAN. Therefore, in this chapter we tried to comprise all the recent information, which help scientist to explore more about in this area of research.

Author details

Rekha Sodani[1], Udit Nandan Mishra[2], Subhash Chand[3], Indu[3], Hirdayesh Anuragi[4], Kailash Chandra[5], Jyoti Chauhan[6], Bandana Bose[6], Vivek Kumar[6], Gopal Shankar Singh[7], Devidutta Lenka[8] and Rajesh Kumar Singhal[3]*

1 College of Agriculture, Nagur, Agriculture University, Jodhpur, Rajasthan, India

2 M.S. Swaminathan School of Agriculture, Centurion University of Technology and Management, Paralakhemundi, Gajpati, Odisha, India

3 ICAR-Indian Grassland and Fodder Research Institute, Jhansi, U.P., India

4 ICAR-Central Agroforestry Research Institute, Jhansi, U.P., India

5 Sri Karan Narendra Agriculture University, Jobner, India

6 Institute of Agriculture Sciences, Banaras Hindu University, U.P., India

7 Institute of Environment and Sustainable Development, Banaras Hindu University, Varanasi, U.P., India

8 Orissa University of Agriculture and Technology, Bhubaneswar, Odisha, India

*Address all correspondence to: rajasinghal151@gmail.com

References

[1] Sharkey TD. Emerging research in plant photosynthesis. Emerging topics in life sciences. 2020; **4**(2): 137-150.

[2] Kami C, Lorrain S, Hornitschek P, Fankhauser C. Light-regulated plant growth and development. In Current topics in developmental biology. CRC Academic Press. 2010: 29-66.

[3] Nozue K, Maloof JN. Diurnal regulation of plant growth. Plant Cell and Environment. 2006; **29**(3): 396-408.

[4] Svechkina A, Portnov BA, Trop T. The impact of artificial light at night on human and ecosystem health: a systematic literature review. Landscape Ecology. 2020: 1-18.

[5] Haim A, Portnov BA. Light pollution as a new risk factor for human breast and prostate cancers. Dordrecht: Springer. 2013. p. 168.

[6] Gaston KJ, Duffy JP, Gaston S, Bennie J, Davies TW. Human alteration of natural light cycles: causes and ecological consequences. Oecologia. 2014; **176**(4): 917-931.

[7] Helm D. The European framework for energy and climate policies. Energy Policy.2014; **64**: 29-35.

[8] Hey MH, DiBiase E, Roach DA, Carr DE, Haynes KJ. Interactions between artificial light at night soil moisture and plant density affect the growth of a perennial wildflower. Oecologia. 2020:1-8.

[9] Falcón J, Torriglia A, Attia D, Viénot F, Gronfier C, Behar-Cohen F et al. Exposure to artificial light at night and the consequences for flora fauna and ecosystems. Frontiers in Neuroscience.2020. p.1183.

[10] Singhal RK, Chauhan J, Jatav HS, Rajput VD, Singh GS, Bose B. Artificial Night Light Alters Ecosystem Services Provided by Biotic Components. Biologia Futura. 2021.

[11] Zhang FS, Wang Y, Wu K, Xu WY, Wu J, Liu JY, Wang XY, Shuai LY. Effects of artificial light at night on foraging behavior and vigilance in a nocturnal rodent. Science of The Total Environment. 2020; p.138271.

[12] Dominoni DM, Kjellberg Jensen J, de Jong M, Visser ME, Spoelstra K. Artificial light at night in interaction with spring temperature modulates timing of reproduction in a passerine bird. Ecological Applications.2020; **30**(3):e02062.

[13] Guetté A, Godet L, Juigner M, Robin M. Worldwide increase in Artificial Light at Night around protected areas and within biodiversity hotspots. Biological conservation. 2018; **223**:97-103.

[14] Chalkias C, Petrakis M, Psiloglou B, Lianou M. Modelling of light pollution in suburban areas using remotely sensed imagery and GIS. Journal of environmental management.2006; **79** (1):57-63.

[15] Falchi F, Cinzano P, Duriscoe D, Kyba CC, Elvidge CD, Baugh K, Portnov BA et al. The new world atlas of artificial night sky brightness. Science advances. 2006; **2**(6).

[16] Linares H, Masana E, Ribas SJ, Garcia-Gil M, Figueras F, Aubé M. Modelling the night sky brightness and light pollution sources of Montsec protected area. Journal of Quantitative Spectroscopy and Radiative Transfer.2018; **217**:178-188.

[17] Cinzano P, Falchi F, Elvidge CD. The first world atlas of the artificial night sky brightness. Monthly Notices of the Royal Astronomical Society.2001; **328**(3):689-707.

[18] Kumar P, Rehman S, Sajjad H, Tripathy BR, Rani M, Singh S. Analyzing trend in artificial light pollution pattern in India using NTL sensor's data. Urban Climate.2019; **27**:272-283.

[19] Bará S. Anthropogenic disruption of the night sky darkness in urban and rural areas. Royal Society Open Science.2016; **3**(10):160541.

[20] Gaston KJ, Bennie J, Davies TW, Hopkins J. The ecological impacts of nighttime light pollution: a mechanistic appraisal. Biological reviews.2013; **88**(4):912-927.

[21] Duca M. Plant Biorhythms. In Plant Physiology. Springer International Publishing. 2015: 231-246.

[22] Bose B, Pant B, Singhal RK, Kumar M, Mondal S. Phytochrome: physiology molecular aspects and sustainable crop production. Emerging Trends of Plant Physiology for Sustainable Crop Production. 2018.

[23] McClung CR. Plant circadian rhythms. The Plant Cell.2006; **18**(4):792-803.

[24] Gwinner E. Circannual rhythms: *endogenous annual clocks in the organization of seasonal processes*. Springer Science & Business Media. 2012; **18**.

[25] Sweeney BM. Rhythmic phenomena in plants. Academic Press. 2013.

[26] Bunning E. Endogenous rhythms in plants. Annual Review of Plant Physiology. 1956; 7(1): 71-90.

[27] Webb AA. The physiology of circadian rhythms in plants. New Phytologist. 2003; **160**(2): 281-303.

[28] Pittendrigh CS. Circadian rhythms and the circadian organization of living systems. In *Cold Spring Harbor symposia on quantitative biology*. Cold Spring Harbor Laboratory Press. 1960; **25**: 159-184.

[29] Leigh EG, Rand AS, Windsor DM. *The ecology of a tropical forest: Seasonal rhythms and long-termchanges* (No. 581.5 L528e). Washington, US: Smithsonian Institution. 1985.

[30] Webb HM, Brown Jr FA. Timing long-cycle physiological rhythms. Physiological reviews 1959; **39**(1):127-161.

[31] Christie JM. Phototropin blue-light receptors. Annual Review of Plant Biology. 2007; 58:21-45.

[32] Millar AJ. Input signals to the plant circadian clock. Journal of Experimental Botany.2014; **55**(395):277-283.

[33] Dunlap JC, Loros JJ, De Coursey P. Chronobiology: Biological Timekeeping. Sunderland MA: Sinauer Associates.2004.

[34] Casal JJ. Phytochromes cryptochromes phototropin: photoreceptor interactions in plants. Photochemistry Photobiology.2000; **71**:1-11.

[35] Fukuda H, Murase H, Tokuda IT. Controlling circadian rhythms by dark-pulse perturbations in Arabidopsis thaliana. Scientific reports.2013;**3**(1):1-7.

[36] Takahashi N, Hirata Y, Aihara K, Mas P. A hierarchical multi-oscillator network orchestrates the Arabidopsis circadian system. Cell.2015; **163**: 148-159.

[37] Hayama R, Coupland G. Shedding light on the circadian clock and the photoperiodic control of flowering. Current Opinion in Plant Biology. 2003; **6**:13-19.

[38] Thomas B, Vince-Prue D. Photoperiodism in plants. London: Academic Press.1996.

[39] Fankhauser C, Staiger D. Photoreceptors in Arabidopsis thaliana: light perception signal transduction and entrainment of the endogenous clock. Planta.2002; **216**:1-16.

[40] Andres F, Coupland G. The genetic basis of flowering responses to seasonal cues. Nature Reviews Genetics. 2012; **13**:627-639.

[41] Suarez-Lopez P, Wheatley K, Robson F, Onouchi H, Valverde F, Coupland G. CONSTANS mediates between the circadian clock and the control of flowering in Arabidopsis. Nature.2001; **410**:1116-1120.

[42] Singhal RK, Kumar V, Kumar M, Bose B. Responses of different light intensities and continue light during dark period on rice (*Oryza sativa* L.) seed germination and seedling development. Journal of Pharmacognosy and Phytochemistry.2019; **8**(4):2602-2609.

[43] Zhang B, Zhang H, Jing Q, Wang J. Light pollution on the growth physiology and chlorophyll fluorescence response of landscape plant perennial ryegrass (*Lolium perenne* L.). Ecological Indicators.2020; **115**:106448.

[44] Kwak MJ, Lee SH, Khaine I, Je SM, Lee TY, You HN et al. Stomatal movements depend on interactions between external night light cue and internal signals activated by rhythmic starch turnover and abscisic acid (ABA) levels at dawn and dusk. Acta Physiologiae Plantarum.2017; **39**(8):162

[45] Kwak MJ, Je SM, Cheng HC, Seo SM, Park JH, Baek SG et al. Night light-adaptation strategies for photosynthetic apparatus in yellow-poplar (Liriodendron tulipifera L.) exposed to artificial night lighting. Forests.2018; **9**(2):74.

[46] Meravi N, Kumar Prajapati S. Effect street light pollution on the photosynthetic efficiency of different plants. Biological Rhythm Research.2020; **51**(1):67-75.

[47] Ffrench-Constant RH, Somers-Yeates R, Bennie J, Economou T, Hodgson D, Spalding A et al. Light pollution is associated with earlier tree budburst across the United Kingdom. Proceedings of the Royal Society B: Biological Sciences.2016; **283**(1833): 20160813.

[48] Bennie J, Davies TW, Cruse D, Bell F, Gaston KJ. Artificial light at night alters grassland vegetation species composition and phenology.2017.

[49] Bennie J, Davies TW, Cruse D, Bell F, Gaston KJ. Artificial light at night alters grassland vegetation species composition and phenology. Journal of applied ecology.2018; **55** (1):442-450.

[50] Viera-Pérez M, Hernández-Calvento L, Hesp PA, Santana-del Pino A. Effects of artificial light on flowering of foredune vegetation. Ecology.2019; **100**(5):e02678.

[51] Yang J, Duan R. The effect of artificial illumination on postponing plant phenology. Applied Ecology and Environmental Research.2019; **17**(6):13289-13296.

[52] Bennie J, Davies TW, Cruse D, Gaston KJ. Ecological effects of artificial light at night on wild plants. Journal of Ecology.2016; **104**(3):611-620.

[53] Škvareninová J, Tuhárska M, Škvarenina J, Babálová D, Slobodníková L, Slobodník B et al. Effects of light pollution on tree phenology in the urban environment. Moravian Geographical Reports.2017; **25**(4):282-290.

[54] Singhal RK, Kumar M, Bose B. Ecophysiological responses of artificial

night light pollution in plants. Russian Journal of Plant Physiology.2019.1-13.

[55] Biesmeijer JC, Roberts SP, Reemer M, Ohlemüller R, Edwards M, Peeters T et al. Parallel declines in pollinators and insect-pollinated plants in Britain and the Netherlands. Science.2006; **313**(5785):351-354.

[56] Potts SG, Imperatriz-Fonseca V, Ngo HT, Aizen MA, Biesmeijer JC, Breeze TD et al. Safeguarding pollinators and their values to human well-being. Nature.2016; **540**(7632):220-229.

[57] Knop E, Zoller L, Ryser R, Gerpe C, Hörler M, Fontaine C. Artificial light at night as a new threat to pollination. Nature. 2017; **548**(7666):206-209.

[58] Macgregor CJ, Pocock MJ, Fox R and Evans DM. Pollination by nocturnal Lepidoptera and the effects of light pollution: a review. Ecological entomology.2015; **40**(3):187-198.

[59] Giavi S, Blösch S, Schuster G, Knop E. Artificial light at night can modify ecosystem functioning beyond the lit area. Scientific reports.2020; **10**(1):1-11.

[60] Singhal RK, Sodani R, Chauhan J, Sharma MK, Yashu BR. Physiological Adaptation and Tolerance Mechanism of Rice (Oryza sativa L.) in Multiple Abiotic Stresses. International Journal of Pure and Applied Biosciences.2017; **5** (3):459-466.

[61] Singhal RK, Kumar V, Kumar S, Choudhary BL. High Light Stress Response and Tolerance Mechanism in Plant. Interdisciplinary journal of Contemporary Research. 2017:4 ISSN: 23938358.

[62] Ayalon I, de Barros Marangoni LF, Benichou JI, Avisar D, Levy O. Red Sea corals under Artificial Light Pollution at Night (ALAN) undergo oxidative stress and photosynthetic impairment. Global change biology.2019;**25** (12):4194-4207.

[63] Abraham H, Scantlebury DM, Zubidat AE. The loss of ecosystem-services emerging from artificial light at night. Chronobiology International. 2019; **36**(2):296-298.

[64] Manfrin A, Singer G, Larsen S, Weiß N, van Grunsven RH, Weiß NS et al. Artificial light at night affects organism flux across ecosystem boundaries and drives community structure in the recipient ecosystem. Frontiers in Environmental Science.2017; **5**:61.

[65] Bureau of Indian Standards. Indian Standard, code of practice for lightning of public thoroughfares, IS 1944-7: 1981 (R2003). New Delhi, India.

[66] The Hindu, https://www.thehindu.com/news/cities/bangalore/tender-of-led-street-lights-project-to-be-cancelled/article30854438.ece. 2018.

[67] Du Jiangtao, Zhang X, King D. 2018. An investigation into the risk of night light pollution in a glazed office building: The effect of shading solutions. Building and Environment 145 DOI: 10.1016/j.buildenv.2018.09.029

[68] 13thMEETING OF THE CONFERENCE OF THE PARTIES, Conventio on migratory species. Agenda Item 26.4 "The guidelaince for light pollution for wildlife including marine turtle, seasbirds and migratory shoebirds". Gandhinagar, India, UNEP/CMS/COP13/Doc.26.4.9.1/Rev.1, 17-22 February 2020.

[69] Schroer S, Hölker F. Light pollution reduction. Handbook of Advanced Lighting Technology; Karlicek, R., Sun, C.-C., Zissis, G., Ma, R., Eds. 2014; 1-17.

[70] Kyba CCM, Hölker F. Do artificially illuminated skies affect biodiversity in nocturnal landscapes? Landscape Ecology. 2013. **28**:1637-1640.

[71] Cinzano P. Technical measures for an effective limitation of the effects of

light pollution. In:Cinzano P (ed) Light pollution and the protection of the night environment. Proceedings of the IDA regional meeting "Venice: let's save the night". ISTIL, Thiene. 2002: 193-205.

[72] Longcore T, Rich C. In: Synthesis Rich C, Longcore T (eds) Ecological consequences of artificial night lighting. Island Press, Washington, DC. 2006: 413-430.

[73] Navara KJ, Nelson RJ. The dark side of light at night: physiological, epidemiological, and ecological consequences. Journal of Pineal Research.2007; **43**:215-224.

[74] Eisenbeis G, Hanel A. Light pollution and the impact of artificial night lighting on insects. In:Eisenbeis G, Hanel A (eds) Ecology of cities and towns. Cambridge University Press, 2009; Cambridge. Books online.

[75] Hölker F, Wolter C, Perkin EK, Tockner K. Light pollution as a biodiversity threat. Trends in Ecology & Evolution. 2010; **25**:681-682.

[76] Bruce-White C, Shardlow M. A review of the impact of artificial light on invertebrates. Buglife – The Invertebrate Conservation Trust, Peterborough, 2011.

[77] Gaston KJ, Davies TW, Bennie J, Hopkins J. Reducing the ecological consequences of nighttime light pollution: options and developments. Journal of Applied Ecology. 2012; 49(**6**):1256-1266

[78] Hölker F. Lichtverschmutzung und die Folgen fur Ökosysteme und Biodiversitat. In: Held M, Hölker F, Jessel B (eds) Schutz der Nacht – Lichtverschmutzung, Biodiversitat und Nachtlandschaft. BfN-Skripten 336, Bonn. 2013; 73-76.

[79] Dick R. Applied scotobiology in luminaire design. Light Research Technology. 2014; 46:50-66.

[80] Elsahragty M, Kim JL. Assessment and strategies to reduce light pollution using geographic information systems. Procedia engineering. 2015; 118: 479-488.

[81] Schroer S, Hölker F. Impact of lighting on flora and fauna. Handbook of Advanced Lighting Technology; Karlicek, R., Sun, C.-C., Zissis, G., Ma, R., Eds .2016: 1-33.

[82] Schroer S, Huggins BJ, Azam C, Hölker F. Working with inadequate tools: Legislative shortcomings in protection against ecological effects of artificial light at night. Sustainability 2020; 12(**6**), 2551.

Chapter 3

Moth Species Caught by Ultraviolet and Visible Light Sources in Connection with Their Wingspan

László Nowinszky, Lionel Hill, János Puskás, Károly Tar and Levente Hufnagel

Abstract

For a long time, researchers have compared light traps operating with different light sources. According to the results, ultraviolet lights often performed better than visible light sources. In the present study, we examine the wingspan of macrolepidoptera species in relation to the catch result of visible (visible) and BL traps in choice and no-choice situations using data from the Hungarian light-trap network. We used the catch data of 19 light-trap stations from 1962 to 1963. Up to 18 stations belonged to the national network and the last one was in Nagytétény. We processed data of 381 species of the 18 light-traps data of the national network and data of 222 species from the light traps of Nagytétény. The data of the wingspan of the different macrolepidoptera species we collected from the websites of UKmoths (http://ukmoths.org.uk/index.php), and Guide to the Butterflies and Moths of Hungary (macrolepidoptera) (http://www.macrolepidoptera.hu). We summarised for each light-trap station and each trap type the number of the macrolepidopteran species and individuals caught from different generations. Then, using the Mann–Whitney test, we checked for species the number of individuals captured by visible and BL traps, and the difference of the level of significance. We summarised the wingspan data of all the 381 species, the more efficient light source for each species in a no-choice situation at multiple sites and for the single site of Nagytétény the more efficient light source for species detected there. The BL trap seems most efficient for operation for plant protecting purposes, despite the fact that their use is far more problematic. Insect species are not only endangered by light trapping but also by the light pollution of urban areas. Our results confirm that the different light sources should incur mortality on different species to differing levels. Such differential mortality from artificial light sources could disturb the balance of life in biological communities.

Keywords: moths, wingspan, light-trap, visible and BL light sources

1. Introduction

For a long time, researchers worldwide have compared light traps operating with different light sources. The results have been diverse although ultraviolet lights often performed better than white light sources. Catching results seem to vary with taxa and with the size of insects; so, it is not easy to assert that one type of light source is best for all occasions. Knowledge of the responses of various insect taxa to artificial light not only has implications for trapping techniques in entomology but also for assessing the impact of artificial light pollution on biological ecosystems.

Researchers have also examined the spectral sensitivity of the insect's eye. Electroretinogram measurements are used to determine the spectral sensitivity of the insect eye. In international literature, several studies are devoted to the results of laboratory measurements carried out on various species. No reports of such experiments are known in Hungary, and data on the most important Hungarian pestilent species are also missing from the international literature on the subject.

Few researchers have examined the relationship of the body or eye size with the selection of light sources.

Different light sources are used in the various types of light traps. The light source determines the running temperature, the colour temperature, and the spectral distribution of the light energy that it emits. Some sources such as black light (BL) emit mainly in the ultraviolet wavelength range (320–400 nm), some such as normal or white light (V) emit in the visible range (400–700 nm), and some such as mercury vapour (Hg) emit across ultraviolet and visible wavelengths (200–600 nm). Some sources emit or are filtered to a narrow range of wavelengths such as particular colours visible to the human eye.

Mikkola [1] established that moths and butterflies (Lepidoptera) and caddis fly species (Trichoptera) have an eye sensitivity that remains practically unchanged in the 350–600 nm spectrum. Its maximum is around 550 nm (green, same as the value of the human eye during daytime). The sensitivity is greatly reduced at about 620 nm (orange-red). McFarlane and Eaton [2] have reported that the responses of Cabbage Looper (*Trichoplusia ni* Hbn.) to monochromatic light stimuli have been investigated by electroretinogram (ERG) and electromyogram (EMG) techniques. The spectral sensitivity curves for male and female Cabbage Loppers show a major peak at 540–550 nm and a minor peak in the ultraviolet range at 360 nm. Agee [3] showed by electroretinogram tests that the sensitivity of eyes of the Bollworm Moth (*Heliothis zea* Boddie) and Tobacco Bollworm (*Heliothis virescens* F.) to 365 nm and 480–575 nm wavelengths light is the highest.

Pappas and Eaton [4] found that the ocelli of the Tobacco Hornworm (*Manduca sexta* L.) are more sensitive to 520 nm light than to 360 nm light stimuli. Similar results are reported by Eguchi et al. [5] about the Sphingid moths. These moths possess the highest peak sensitivity at 540 nm.

Gui et al. [6] reported that the colours on which comparable data are available to arrange themselves in order of least to most attractiveness to insects as red, yellow, white, and blue. From tests of Taylor and Deay [7], it appears that the maximum attractiveness for the European Corn Borer (*Ostrinia nubilalis* Hbn.) is in the near-ultraviolet region between 320 and 380 nm.

Many researchers found that ultraviolet or black light was most effective in catching insects but for some taxa, a combination of ultraviolet and visible light was more effective while a few taxa were best trapped by using visible light alone.

In a comparative experiment, Frost [8] found that black light attracted almost all taxa of insects more than white light. The exceptions were the Miridae and Chrysopidae, which preferred white light. Belton and Kempster [9] (1963) caught more Noctuidae and Geometridae with a BL fluorescent tube than with the normal or cold white light (N). Sifter [10] examined the swarming of the Chestnut Weevil (*Curculio elephas* Gyllenhal, Coleoptera: Curculionidae) by using visible and BL light traps. The body length of this beetle is only 6–9 mm. The normal light trap did not catch a single specimen but the BL one was suitable for investigation of swarming. According to Bürgés et al. [11], those families (Geometridae, Sphingidae, Notodontidae, Arctiidae, and Noctuidae) that are rich in species fly to both normal and BL light traps, but the BL traps catch significantly more species and more specimens of many species [12] studied the efficiency of catching useful or beneficial insects by using different light sources. The Coccinellidae (Coleoptera) species preferred BL, the *Ophion* sp. (Hymenoptera: Ichneumonidae) preferred blue BL while *Chrysopa* spp. (Neuroptera: Chrysopidae) was trapped equally well with white and BL light, while every source of light had the same impact on some broad damsel bugs (Hemiptera: Nabidae) and the lacewings, *Hemerobius* spp. (Neuroptera: Hemerobiidae). Comparative studies [13] found each of several Microlepidoptera species was more effectively collected in BL traps than normal ones. In our earlier study [14], we compared in two light-trap stations, the composition of species of five macrolepidopteran families from the material of normal and BL light traps by applying the Sorensen index. The results are the follows: Geometridae: 0.607 and 0.518; Sphingidae: 0.750 and 0.500; Notodontidae: 0.444 and 0.429; Arctiidae: 0.714 and 0.609; Noctuidae: 0.608 and 0.527.

Some authors found ultraviolet light more effective than particular wavelengths of visible light in trapping insects. In the test of Day and Reid [15], the 15 W fluorescent BL lamps were more efficient for capturing *Conoderus falli* Lane (Coleoptera: Elateridae) than similar yellow sources. Teel et al. [16] perceived the maximum sensitivity of the eye of Hickory Shuckworm (*Laspeyresia caryana* Fitch.) at 365 nm and 515 nm. At these two values, six times as many individuals responded to the near-ultraviolet light than to the green light. Skuhravý et al. [17] found a BL trap much more effective than either yellow, green, or red light in collecting the Saddle Gall Midge (*Haplodiplosis marginata* von Roser) (Diptera: Cecidomydae).

Some authors found a combination of ultraviolet and visible wavelengths to be most effective. Cleve [18] found an ultraviolet fluorescent lamp (BL) that was very attractive to insects if it illuminated a white sheet. Similarly, Belton and Kempster [19] verified the results of their laboratory measurements of eye sensitivity by the test of light-trap collecting. They caught the highest number of insects with lamps emitting both BL and visible light. The catch dwindled when they used BL alone while visible light alone produced an even poorer result. A striking contradiction was found, however, for the six most important insect groups (Coleoptera, Trichoptera, Lepidoptera, Brachycera, and Nematocera Ichneumonoidea) in terms of sensitivity and attractive lighting effect. These insects' eyes were more sensitive to the yellow light but the attractive effect was the opposite.

Some authors found visible light to be more effective than ultraviolet light in trapping certain insects. Jászainé [20] analysed the catching results of Common Meadow Bug (*Exolygus pratensis* Wagner) (Heteroptera: Miridae) in normal (V) and ultraviolet light traps (BL) to find the former caught more individuals. Other taxa showing a greater attraction to regular light include some fruit flies [21] virus vector cicadae (*Laodelphax striatella* (Fallén) and *Javesella pellucida* (Fabr., Homoptera, Areopidae) [22] European Grapevine Moth (*Lobesia botrana* Den. et Schiff.) and Vine Moth (*Eupoecilia ambiguella* Hbn. [23].

Some authors included light sources, such as mercury and sodium in their experiments. For [24] the standard light trap caught only a few specimens of the Eurasian Hemp Moths (*Grapholita delineana* Walker) while a HgLS light source caught many of these moths. The wingspan of the Eurasian Hemp Moths is 10–14 mm.

Blomberg et al. [25] compared two types of light trap catch results. One of them was the so-called blended light trap containing a 160 W Tungsram mercury fluorescent lamp emitting ultraviolet and visible light. The BL was provided with a 125 W Philips HPW lamp. The mercury fluorescent lamp caught twice as many moths of the macrolepidoptera (families Geometridae and Noctuidae), and the microlepidopteran species as the BL trap.

According to Gál et al. and Bürgés [26–28] for light trapping of Chestnut Weevil (*Curculio elephas* Gyllenhal) and Acorn Moth (*Cydia splendana* Hbn.) the most effective light source is the mercury vapour lamp (HgW). Traps with visible or BL lamps achieved comparable catches to each other but less than the mercury source, which produces both ultraviolet and visible light.

Extremely valuable conclusions follow from a series of experiments by Járfás et al., Járfás and Tóth [29, 30] in which catch results yielded by 125 W (HgVE 27) ultraviolet, 125 W (HgLSE27) mercury vapour, 100 W (OHP 220–230 VAO) krypton, 100 W (F_3) 50 cm neon, 250 W (E 279043 IMP) infraruby, and 50 cm germicidal lamps were compared. Silver Y moths (*Autographa gamma* L.), Pine Chafers (*Polyphylla fullo* L.), Vine Chafers (*Anomala vitis* Fabr.), and Scarab Beetles (*Anoxia orientalis* Kryniczky) flew to the mercury vapour lamps in the highest numbers, while infraruby light proved to be practically unsuitable for trapping. Járfás published further results of his experiments on different moth species. Most suitable for catching was the mercury lamp (HgW) ahead of BL which was better than visible or visible light for the Silver Y (*A. gamma* L.) [29], the Codling Moth (*Cydia pomonella* L.) [31], the Pea Podborer (*Etiella zinckenella* Tr.) [32] and the Beet Webworm (*Loxostege sticticalis* L.) [33]. Also, Járfás [34] reported that the Apple Peel Tortrix (*Adoxophyes reticulana* Hbn.), the Pear Moth (*Laspeyresia pyrivora* Pan.), and the Plum Fruit Moth (*Grapholita funebrana* Tr.) can be best caught with the mercury vapour lamp (HgW) but for the Strawberry Tortricid (*Pandemis dumetana* Tr.) and the Dark Fruit-tree Tortrix (*Pandemis heparana* Den. et Schiff.) the visible light bulb was most effective. Similarly, the European Corn Borer (*O. nubilalis* Hbn.) was collected in the HgW traps more successfully than in the visible and the BL traps [35].

Wallner et al. [36] carried out experiments with three lymantriid species in the Russian Far East. They caught significantly more moths of all three species using fluorescent black light than either phosphor mercury or high-pressure sodium lamps. The species were Gipsy Moth (*Lymantria dispar* L.), Nun Moth (*Lymantria monaca* L.), and the Pink Gipsy Moth (*Lymantria matura* Moore).

Fayle et al. [37] compared three types of Robinson light traps equipped with 125 W mercury bulb, which emits visible and ultraviolet light. One of these light sources included materials that absorb visible light; so, this lamp was an ultraviolet or BL type trap. The fewest moths were caught by the BL trap. Barghini [38] tested four light sources. Most insects were caught using the high-pressure mercury lamp (Hg). A further order was as follows—high-pressure sodium (Na) without a BL filter and the same type with BL filter.

In the last decade, most researchers found a connection between the body size of the insects, expressed as body weight, eye size or wingspan, and their light sensitivity. Taxa with larger eyes and wingspan have higher light sensitivity than those with smaller eyes. Over the last decade, published studies supported the finding that the vision of insects with greater body weight is more sensitive than the smaller species. Such a statement was published concerning desert ants (*Cataglyphis*) [39]; pollen foraging bees, (Apoidea) [40] the bumblebees (*Bombus terrestris* L.) [41, 42]; the nymphalid butterflies (Nymphalidae) [43]. Moser et al. [44] found a connection between the size of the eyes of 10 *Atta* species (Hymenoptera: Formicidae) and the time of nuptial flight using the digital photograph method. The diameter of compound eyes of the night flying species was significantly larger. Yack et al. [45] reported similar results in the *Macrosoma eliconiaria* Walker (Lepidoptera: Hedyloidea) species.

Experiments of Kino and Oshima [46] suggest that moth and butterfly emanations could cause allergy-induced bronchial asthma in certain patients. Since moths are attracted readily to artificial lights and often fly into houses, these insects are especially suspect as important factors in extrinsic asthma. Barghini and Medeiros [47] (2010) assumed that in developing countries, the growing light pollution will affect the spread of vector-borne human diseases as well.

van Langevelde et al. [48] established that artificial light with smaller wavelengths attracted more individuals and greater specific diversity of insects than light with larger wavelengths. The attraction was correlated with the body mass, wingspan, and eye size of moths. The size-dependent response to artificial light sources is likely to distort the ecosystems if it generates selective mortality.

In the above-mentioned studies, the catch coming from parallel operated regular and BL light traps offered a unique possibility to answer the following questions.

- Is there a significant difference by species and families between the catch yielded by the two types of traps?
- Which of the two light sources is more suitable for trapping what species?
- Are there any species that can only be collected by one of the two types of light sources?
- Does either of the two types indicate the presence of more species than the other?
- To what extent do the materials yielded by the two types of traps at the same observation site differ in their composition by species?

In the present study, we examined the wingspan of macrolepidopteran species in relation to the catch result of visible and BL traps in choice and no-choice situations using data from the Hungarian light-trap network.

2. Material

To compare the differences in the practical use of visible and BL light traps, from 1962, the Hungarian Plant Protection Research Institute at Keszthely experimented with the parallel operation of two light traps, one running on a visible bulb producing mainly visible light and the other outfitted with BL light-emitting mainly ultraviolet light. Also in 1962, the Plant Protection Service, in its turn, added a BL light trap in Nagytétény to the ones running on visible light, and equipped all its county plant protection stations also with BL traps in 1963. The national network of parallel operated visible and BL light traps opened up the possibility to a wide-scale examination of the results and usefulness of collecting with the two types. Most valuable information was provided by the light traps at Nagytétény where regular and BL traps were placed at a mere 10 metres distance from one another. The proximity of the two traps meant an identity of the microclimate, vegetation, and the distance from the habitats of the various species and so the insects were practically offered the choice of the two different light sources. At other sites, the visible and BL traps were separated by a distance greater than their likely radius of effect and so did not offer a choice situation to insects.

The visible and BL light traps operated in the following cities and villages:

Baj (47°38′N, 18°21′E)	Mikepércs (47°26′N, 21°37′E)
Csopak (45°58′N, 17°55′E)	Miskolc (48°51′N, 20°46′E)
Fácánkert (46°26′N, 18°44′E)	Nagytétény (47°38′N, 18°97′E)
Gyöngyös (47°46′N, 19°55′E)	Pacsa (46°43′N, 17°09′E)
Győr-Kismegyer (47°39′N, 17°39′E)	Szederkény (45°59′N, 18°27′E)
Hódmezővásárhely (46°25′N, 20°19′E)	Tanakajd (47°11′N, 16°44′E)
Kaposvár (46°22′N, 17°46′E)	Tarhos (46°48′N, 21°12′E)
Kállósemjén (47°51′N, 21°55′E)	Tass (47°12′N, 19°20′E)
Kenderes (47°13′N, 20°45′E)	Velence (47°14′N, 18°38′E)
Keszthely (46°46′N, 17°15′E)	

The complete macrolepidopteran material of above-listed light traps was processed in our work. We processed data of 381 species of the 18 light-traps data of the national network and data of 222 species from the light traps of Nagytétény.

The data of the wingspan of the different Macrolepidoptera species we collected from the websites of UKmoths (http://ukmoths.org.uk/index.php), and Guide to the Butterflies and Moths of Hungary (macrolepidoptera) (http://www.macrolepidoptera.hu).

3. Methods

We summarise for each light-trap station and each trap type the number of the macrolepidopteran species and individuals caught from different generations but did not separate the individuals into generations. Then, using the Mann–Whitney test, we checked for species the number of individuals captured by visible and BL traps, and the difference of the level of significance. The theoretical bases of the test and its application were shown by Hajtman et al. [49, 50] in detail. We created a common sample in the course of the procedure, which included all of the observation sites. The element number of the sample is twice the number of observation sites (because two traps were in operation at every station), at which one of the traps revealed the presence of a species. We sum it up by segregating the numbers of individuals in the unified sample. We compared these values with the table value to determine the difference and its level of significance.

Particular attention was paid to the comparison of catches at Nagytétény in the visible and BL traps which were in close proximity, with identical micro-climate, vegetation, and habitat, so that the moths could choose between different light sources at one place.

In the taxonomic sequence, we tabulate all species for wingspan and preferred type of light trap. We separate in this table the light traps of the nationwide network (no-choice situations) from the light traps at Nagytétény (choice situation).

For graphical analysis, we arranged in ascending order, regardless of their taxonomic place, all the species collected both by the national light-trap network, as well as the Nagytétény traps according to the wingspan of insects. We calculated the percentages of species caught by BL and visible traps in relation to the sum of data of the network and also Nagytétény. We calculated the approaching functions of the curves.

The approximate curve is the so-called logistic curve:

$$y = \frac{k}{(1 + e^{b_0 + b_1 x})}$$

where "k" is the saturation value [51]. In our case, $k = 100$, because the elements of samples are in percentage. So, we must not estimate the value of k from the samples. In this way, the values of b_0 and b_1 can be determined by linear regression of transformed data. The estimated values of these constants are: $b_0 = 3.19$, $b_1 = -0.151$.

The value of the correlation index can be determined from the relationship:

$$i_{xy} = \sqrt{1 - \frac{s_r^2}{s_y^2}}$$

where s_r^2 is the residual variance, s_y^2 is the variance of the independent variable? In our case: $i_{xy} = 0.956$.

We depicted their number as the species in the function of the wingspan, that BL and the visible light traps collected it in an equal proportion. We made use of the middle values of the extreme values in all cases. We examined in Ref. to the families Sphingidae, Geometridae, Notodontidae, Erebidae, and Noctuidae whether the number of species collected effectively by the visible or BL traps differed? We also looked for species that cannot be detected in the two results (visible versus BL) in significant differences despite the number of traps being sufficient to determine significant differences.

4. Results and discussion

We summarise in **Table 1** that the wingspan data of all the 378 species, the more efficient light source for each species in a no-choice situation at multiple sites and for the single site of Nagytétény the more efficient light source for species detected there.

We established from the material of the national light-trap network that the BL traps are unquestionably more efficient in collecting several species of the Sphingidae, Notodontidae, and Noctuidae. Several species of the Geometridae and Erebidae families fly to BL and visible traps in equal numbers. However, at Nagytétény, the species of the latter two families clearly flew much more frequently into the BL trap. None of the five families include species that could be captured only by one or the other type of trap.

Figure 1 shows that at no-choice sites, such as the national network traps, 30 mm wingspan is approximately the limit below which some species can be trapped more effectively by using the visible trap rather than the BL type. Above 35 mm wingspan, the catch of the BL approaches 100%. At Nagytétény, however, where the visible and the BL traps were placed so close together that the moths could see both at the same time, even the moths having the smallest wingspan were caught more than 60% by BL trap (**Figure 2**). These results agree broadly with the previous literature although they do not address mercury light sources, which emit light in both BL and V ranges.

Figure 3 shows that the number of the species collected in nearly equal proportions by visible and BL traps significantly declines with increasing wingspan.

No.	Scientific names of species	A	B	C	D
Drepanidae (Average of wingspan is 31.6 mm)					
1	*Watsonalla binaria* Hfn.	24	10	E	—
2	*Drepana falcataria* L.	31	6	E	—
3	*Sabra harpagula* Esp.	30	5	E	—
4	*Cilix glaucata* Scop.	20	20	E	E
5	*Asphalia ruficollis* Den. et Schiff.	36	4	E	—
6	*Habrosyne pyrithoides* Hfn.	37	5	E	—
7	*Tethea ocularis* Hbn.	35	4	BL	—
8	*Tethea or* Den. et Schiff.	40	7	E	—
Lasiocampidae (Average of wingspan is 42.2 mm)					
9	*Poecilocampa populi* L.	37	6	E	—
10	*Trichiura crataegi* L.	27	5	E	—
11	*Malacosoma neustria* L.	30	12	E	—
12	*Lasiocampa trifolii* Den. et Schiff.	47	6	E	BL
13	*Odonestis pruni* L.	40	17	BL	—
14	*Macrothylacia rubi* L.	52	11	E	—
15	*Phyllodesma ilicifolia* L.	35	11	BL	—
16	*Gastropacha quercifolia* L.	70	17	BL	E
Saturniidae (Average of wingspan is 82.5 mm					
17	*Saturnia pyri* Den. et Schiff.	115	7	BL	—
18	*Saturnia pavonia* L.	50	5	BL	—
Sphingidae (Average of wingspan is 82.5 mm					
19	*Mimas tiliae* L.	67	12	BL	BL
20	*Smerinthus ocellata* L.	75	21	BL	BL
21	*Laothoe populi* L.	77	17	BL	V
22	*Marumba quercus* Den. et Schif.	100	5	E	—
23	*Agrius convolvuli* L.	100	14	BL	—
24	*Sphinx ligustri* L.	105	19	BL	E
25	*Sphinx pinastri* L.	77	11	BL	—
26	*Macroglossum stellatarum* L.	45	5	E	—
27	*Deilephila elpenor* L.	53	14	BL	—
28	*Deilephila porcellus* L.	43	14	BL	BL
29	*Hyles euphorbiae* L.	65	21	BL	BL
Geometridae (Average of wingspan is 26.1 mm)					
30	*Rhodostrophia vibicaria* Clerck	27	16	E	BL
31	*Idaea rufaria* Hbn.	13	6	E	V
32	*Idaea serpentata* Hfn.	22	5	E	—
33	*Idaea aureolaria* Den. et Schiff.	11	4	BL	—
34	*Idaea muricata* Hfn.	19	8	E	—
35	*Idaea rusticata* Den & Schiff.	20	17	E	BL

No.	Scientific names of species	A	B	C	D
36	*Idaea obsoletaria* Rambur	22	4	V	—
37	*Idaea fuscovenosa* Goeze.	20	12	E	BL
38	*Idaea humiliata* Hfn.	20	12	V	V
39	*Idaea politaria* Hbn.	16	5	V	—
40	*Idaea seriata* Schrk.	20	5	E	BL
41	*Idaea dimidiata* Hfn.	16	17	V	V
42	*Idaea nitidata* H.-Sch.	20	4	E	BL
43	*Idaea aversata* L.	26	16	E	BL
44	*Idaea degeneraria* Hbn.	28	7	E	BL
45	*Idaea straminata* Brkh.	30	10	E	BL
46	*Scopula immorata* L.	23	17	E	BL
47	*Scopula nigropunctata* Hfn.	31	4	E	—
48	*Scopula virgulata* Den. et Schiff.	20	20	V	E
49	*Scopula ornata* Scop.	22	11	V	V
50	*Scopula rubiginata* Hfn.	18	19	E	E
51	*Scopula marginepunctata* Goeze	26	18	E	E
52	*Scopula immutata* L.	25	17	V	E
53	*Scopula rubiginata* Hfn.	18	19	E	E
54	*Scopula marginepunctata* Goeze	26	18	E	E
55	*Scopula flaccidaria* Zeller	21	14	E	—
56	*Scopula corrivalaria* Kretschm.	20	5	E	—
57	*Scopula incanata* L.	26	7	E	—
58	*Timandra comae* Schmidt	25	22	V	E
59	*Cyclophora annularia* Fabr.	20	15	E	BL
60	*Cyclophora ruficiliaria* H.-Sch.	27	4	E	—
61	*Cyclophora punctaria* L.	22	14	E	—
62	*Cyclophora linearia* Hbn.	29	8	BL	—
63	*Philbalapteryx virgata* Hfn.	23	7	E	BL
64	*Lythria purpuraria* L.	24	15	E	BL
65	*Orthonama vittata* Bkh.	24	7	E	—
66	*Nycterosea obstipata* Fabr.	19	16	E	BL
67	*Xanthorrhoe fluctuata* L.	21	20	E	BL
68	*Xanthorrhoe spadicearia* Den. et Schiff.	25	4	E	—
69	*Xanthorrhoe ferrugata* Clerck	20	16	V	V
70	*Catarhoe cuculata* Hfn.	24	4	E	—
71	*Catarhoe rubidata* Den. et Schiff.	28	6	V	V
72	*Costaconvexa polygrammata* Bkh.	26	7	E	—
73	*Epirrhoe alternata* Müller	22	15	E	BL
74	*Epirrhoe galiata* Den. et Schiff.	30	5	E	—
75	*Pelurga comitata* L.	27	13	E	V

No.	Scientific names of species	A	B	C	D
76	*Gandaritis pyraliata* Den. et Schiff.	30	4	E	—
77	*Operophthera brumata* L.	25	11	V	—
78	*Philereme vetulata* Den. et Schiff.	27	9	E	BL
79	*Perizoma alchemillata* L.	16	10	BL	V
80	*Gymnoscelis rufifasciata* Haw.	17	5	E	BL
81	*Pasiphila rectangulata* L.	17	5	E	—
82	*Eupithecia linariata* Den. et Schiff.	13	14	E	V
83	*Eupithecia simpliciata* Haw.	22	12	E	BL
84	*Eupithecia innotata* Hfn.	21	4	E	BL
85	*Eupithecia centaureata* Den. et Schiff.	18	22	E	BL
86	*Eupithecia vulgata* Haw.	16	7	E	—
87	*Eupithecia millefoliata* Rossler	21	8	V	—
88	*Aplocera plagiata* L.	40	12	E	BL
89	*Lithostege griseata* Den. et Schiff.	29	11	E	BL
90	*Lithostege farinata* Hfn.	31	19	E	E
91	*Abraxas grossulariata* L.	37	5	E	—
92	*Lomaspilis marginata* L.	34	11	E	—
93	*Ligdia adustata* Den. et Schiff.	22	15	E	BL
94	*Stegania dilectaria* Hbn.	21	9	E	—
95	*Macaria alternata* Den. et Schiff.	24	17	E	E
96	*Macaria artesiaria* Den. et Schiff.	26	6	E	—
97	*Narraga tessularia* Metzner	15	6	E	—
98	*Chiasmia clathrata* L.	23	22	E	BL
99	*Epione repandaria* Hfn.	27	7	E	—
100	*Angerona prunaria* L.	40	9	E	—
101	*Ennomos autumnaria* Werneburg	45	16	E	BL
102	*Ennomos fuscantaria* Haw.	37	11	BL	—
103	*Ennomos erosaria* Den. et Schiff.	32	12	BL	—
104	*Selenia lunaria* Den. et Schiff.	39	16	E	V
105	*Artiora evonymaria* Den. et Schiff.	29	4	V	V
106	*Crocallis elinguaria* L.	36	5	E	BL
107	*Colotois pennaria* L.	40	8	E	—
108	*Alsophila aescularia* Den. et Schiff.	30	4	E	—
109	*Ascotis selenaria* Den. et Schiff.	43	21	E	BL
110	*Lycia hirtaria* Clerck	40	9	E	BL
111	*Biston betularia* L.	47	11	BL	BL
112	*Agriopsis bajaria* Den. et Schiff.	29	7	E	—
113	*Therapis flavicaria* Den. et Schiff.	29	5	V	—
114	*Erannis defoliaria* Clerck.	35	7	E	BL
115	*Peribatodes rhomboidaria* Den. et Schiff.	34	13	E	BL

No.	Scientific names of species	A	B	C	D
116	*Cleora cinctaria* Den. et Schiff.	31	6	E	—
117	*Agriopis aurantiaria* Hbn.	31	20	E	BL
118	*Ectropis crepuscularia* L.	35	20	V	BL
119	*Elicrinia trinotata* Metzner	13	6	E	—
120	*Heliomata glarearia* Den. et Schiff.	18	14	E	BL
121	*Synopsia sociaria* Hbn.	36	5	E	—
122	*Aethalura punctulata* Den. et Schiff.	32	5	E	—
123	*Ematurga atomaria* L.	26	17	E	BL
124	*Bupalus piniaria* L.	32	4	BL	—
125	*Cabera pusaria* L.	26	10	E	—
126	*Cabera exanthemata* Scop.	32	15	E	BL
127	*C. exanthemata* Scop.	32	15	E	BL
128	*Lomographa temerata* Den. et Schiff.	24	4	E	—
129	*Tephrina arenacearia* Den. et Schiff.	25	22	E	BL
130	*Tephrina murinaria* Den. et Schiff.	28	11	E	BL
131	*Thetidia smaragdaria* Prout	35	15	E	—
132	*Phaiogramma etruscaria* Zeller	19	9	V	V
133	*Hemistola chrysoprasaria* Esp.	30	10	E	—
134	*Thalera fimbrialis* Scop.	27	16	E	—
135	*Chlorissa cloraria* Hbn.	15	6	E	—
136	*Chlorissa viridata* L.	25	20	V	V
Notodontidae (Average of wingspan is 39.7 mm)					
137	*Thaumetopoea processionea* L.	30	8	E	BL
138	*Cerura vinula* L.	57	7	BL	—
139	*Furcula furcula* Clerk	31	12	BL	BL
140	*Furcula bifida* Brahm	40	16	BL	BL
141	*Drymonia dodonea* Den. et Schiff.	35	6	E	—
142	*Drymonia querna* Fabr.	41	7	E	—
143	*Drymonia ruficornis* Hfn.	37	4	E	—
144	*Notodonta dromedarius* L.	37	5	E	—
145	*Notodonta ziczac* L.	47	17	BL	BL
146	*Notodonta tritophus* Den. et Schiff.	50	6	BL	—
147	*Pheosia tremula* Clerk	50	14	BL	BL
148	*Pterostoma palpina* Clerck	45	20	V	N
149	*Ptilodon capucina* L.	37	5	E	—
150	*Ptilophora plumigera* Den. et Schiff.	38	6	E	BL
151	*Spatalia argentina* Den. et Schiff.	37	11	BL	—
152	*Phalera bucephala* L.	48	17	BL	BL
153	*Gluphisia crenata* Bray	35	13	E	—
154	*Clostera curtula* L.	31	14	V	BL

No.	Scientific names of species	A	B	C	D
155	*Clostera pigra* Hfn.	24	9	E	N
156	*Clostera anastomosis* L.	35	14	E	BL
Erebidae (Average of wingspan is 37,7 mm)					
157	*Scoliopteryx libatrix* L.	42	11	E	V
158	*Rivula sericealis* Scop.	20	21	E	E
159	*Hypena proboscidalis* L.	31	4	E	—
160	*Hypena rostralis* L.	30	12	E	E
161	*Leucoma salicis* L.	43	7	E	—
162	*Lymantria dispar* L.	43	18	BL	BL
163	*Ocneria rubea* Den. et Schiff.	39	6	E	E
164	*Euproctis chrysorrhoea* L.	39	15	E	—
165	*Euproctis similis* Fuessly	31	5	E	—
166	*Calliteara pudibunda* L.	50	7	E	—
167	*Orgya antiqua* L.	27	7	BL	BL
168	*Hyphantria cunea* Drury	38	21	E	BL
169	*Spilosoma lutea* Hfn.	34	18	E	BL
170	*Spilosoma lubricipeda* L.	41	20	E	E
171	*Spilosoma urticae* Esp.	42	18	E	E
172	*Diaphora mendica* Clerck	33	8	E	BL
173	*Diacrisia sannio* L.	42	14	E	—
174	*Phragmatobia fuliginosa* L.	32	22	BL	BL
175	*Phragmatobia lucufer* Den. et Schiff..	37	7	E	BL
176	*Arctia caja* L.	55	21	BL	BL
177	*Arctia villica* L.	52	14	BL	V
178	*Ocnogyna parasita* Hbn.	32	5	E	—
179	*Chelis maculosa* Gerning	33	11	E	BL
180	*Miltochrista miniata* Forster	25	4	E	—
181	*Pelosia muscerda* Hfn.	26	7	E	—
182	*Thumatha senex* Hbn.	17	12	E	—
183	*Pelosia obtusa* H-Sch.	25	8	E	BL
184	*Lithosia quadra* L.	45	12	E	BL
185	*Eilema lurideola* Zincken	31	4	E	—
186	*Eilema complana* L.	31	15	BL	BL
187	*Eilema palliatella* Scop.	34	7	BL	BL
188	*Dysauxes ancilla* L.	23	12	E	—
189	*Eilema pygmaeola* Doubleday	26	14	E	BL
190	*Eilema sororcula* Hfn.	28	5	BL	—
191	*Paracolax tristalis* Fabr.	31	11	E	—
192	*Herminia tarsicrinalis* Knoch.	30	8	E	V
193	*Polypogon tentacularia* L.	25	4	E	—

No.	Scientific names of species	A	B	C	D
194	*Zanclognatha lunalis* Scop.	34	7	V	—
195	*Simplicia rectalis* Ev.	29	6	E	BL
196	*Schrankia costaestrigalis* Steph.	19	5	V	—
197	*Lygephila craccae* Den. et Schiff.	43	7	BL	BL
198	*Phytometra viridaria* Cl.	19	9	E	BL
199	*Colobochyla salicalis* Den. et Schiff.	28	7	E	—
200	*Catocala elocata* Esp.	75	10	BL	BL
201	Euclidia glyphica L.	27	14	E	E
Noctuidae (Average of wingspan is 34,73 mm)					
202	*Eublemma purpurina* Den. et Schiff.	25	18	BL	BL
203	*Abrostola triplasia* L.	30	10	E	—
204	*Abrostola trigemina* Werneburg	37	10	E	BL
205	*Autographa gamma* L.	40	21	BL	BL
206	*Macdunnoughia confusa* Steph.	35	21	E	BL
207	*Diachrysia chrysitis* L.	31	21	BL	E
208	*Plusia festucae* L.	38	12	BL	—
209	*Deltote pygarga* Hfn.	21	9	E	—
210	*Deltote deceptoria* Scop.	24	4	E	—
211	*Deltote uncula* Clerck	21	13	E	—
212	*Deltote bankiana* Fabr.	26	11	E	—
213	*Acontia lucida* Hfn.	28	22	BL	BL
214	*Acontia trabealis* Scop.	19	22	E	BL
215	*Odice arcuinna* Hbn.	27	4	E	BL
216	*Aedia funesta* Esp.	32	20	BL	E
217	*Tyta luctuosa* Den. et Schiff.	23	22	E	BL
218	*Colocasia coryli* L.	34	10	E	—
219	*Diloba caeruleocephala* L.	35	15	E	—
220	*Symira albovenosa* Goeze.	38	10	E	V
221	*Symira nervosa* Den. et Schiff.	32	7	E	—
222	*Acronicta tridens* Den. et Schiff.	40	16	BL	BL
223	*Acronicta psi* L.	40	7	E	BL
224	*Acronicta aceris* L.	45	4	BL	BL
225	*Acronicta rumicis* L.	34	21	BL	BL
226	*Acronicta megacephala* Den. et Schiff.	42	20	BL	BL
227	*Oxycesta geographica* Fabr.	25	4	V	V
228	*Craniophora ligustri* Den. et Schiff.	38	10	BL	—
229	*Cucullia umbratica* L.	47	22	BL	BL
230	*Cucullia chamomillae* Den. et Schiff.	41	4	E	BL
231	*Cucullia lactucae* Den. et Schiff.	48	5	BL	—
232	*Cucullia fraudatrix* Ev.	38	6	E	—

No.	Scientific names of species	A	B	C	D
233	*Lamprosticta culta* Den. et Schiff.	42	4	BL	BL
234	*Ammoconia caecimacula* Den. et Schiff.	42	10	BL	BL
235	*Calophasia lunula* Hfn.	29	18	E	BL
236	*Amphipyra pyramidea* L.	46	5	BL	BL
237	*Amphipyra livida* Den. et Schiff.	42	8	E	BL
238	*Amphipyra tragopoginis* Clerck	35	17	BL	BL
239	*Asteroscopus sphinx* Hfn.	44	11	E	—
240	*Allophyes oxyacanthae* L.	42	7	E	BL
241	*Pyrrhia umbra* Hfn.	31	15	E	BL
242	*Protoschinia scutosa* Den. et Schiff.	33	5	E	—
243	*Heliothis viriplaca* Hfn.	33	21	BL	BL
244	*Heliothis maritima* Graslin	33	22	BL	BL
245	*Periphanes delphinii* L.	36	19	BL	BL
246	*Acosmetia caliginosa* Hbn.	27	10	V	E
247	*Eucarta virgo* Tr.	35	13	E	—
248	*Cryphia algae* Fabr.	27	4	BL	—
249	*Cryphia raptricula* Den. et Schiff.	32	10	BL	—
250	Pseudeustrotia candidula Den. et Schiff.	22	21	E	BL
251	*Spodoptera exigua* Hbn.	29	12	BL	BL
252	*Elaphria venustula* Hbn.	21	7	E	—
253	*Episema glaucina* Esp.	36	9	E	E
254	*Episema tersa* Den. et Schiff.	36	11	BL	BL
255	*Caradrina morpheus* Hfn.	35	17	E	BL
256	*Platyperigea kadenii* Freyer	30	8	BL	BL
257	*Paradrina clavipalpis* Scop.	30	21	BL	BL
258	*Hoplodrina respersa* Hbn.	30	8	BL	V
259	*Hoplodrina alsines* Brahm.	31	17	BL	BL
260	*Hoplodrina respersa* Den. et Schiff.	31	4	E	—
261	*Hoplodrina blanda* Den. et Schiff.	33	14	BL	BL
262	*Hoplodrina ambigua* Den. et Schiff.	33	19	BL	BL
263	*Chilodes maritimus* Tauscher	33	7	E	BL
264	*Charanyca trigrammica* Hfn.	37	16	E	BL
265	*Athetis gluteosa* Tr.	25	19	E	BL
266	*Athetis furvula* Hbn.	20	11	E	—
267	*Dypterygia scabriuscula* L.	34	13	E	BL
268	*Trachea atriplicis* L.	40	12	E	—
269	*Actinotia polyodon* Clerck	33	5	E	—
270	*Phlogophora meticulosa* L.	47	12	BL	BL
271	*Euplexia lucipara* L.	29	7	E	—
272	*Gortyna flavago* Den. et Schiff.	37	9	E	—

No.	Scientific names of species	A	B	C	D
273	*Hydraecia micacea* Esp.	36	5	E	BL
274	*Luperina testacea* Den. et Schiff.	32	22	E	BL
275	*Rhizedra lutosa* Hbn.	46	18	BL	BL
276	*Nonagria typhae* Thnbg.	47	6	E	BL
277	*Archanara geminipuncta* Haw.	29	5	E	BL
278	*Archanara dissoluta* Tr.	30	4	E	—
279	*Denticucullus pygmina* Haw.	26	10	E	BL
280	*Photedes fluxa* Hbn.	28	9	E	BL
281	*Globia sparganii* Esp.	36	8	E	—
282	*Globia algae* Esp.	38	7	E	—
283	*Apamea anceps* Den. et Schiff.	37	15	BL	—
284	*Apamea sordens* Hfn.	38	16	E	E
285	*Apamea monoglypha* Hfn.	50	13	BL	BL
286	*Apamea sublustris* Esp.	42	5	E	V
287	*Mesapamea secalis* L.	28	7	BL	—
288	*Mesoligia furuncula* Den. et Schiff.	25	9	E	BL
289	*Oligia latruncula* Den. et Schiff.	25	19	E	E
290	*Oligia strigilis* L.	23	17	BL	E
291	*Xanthia gilvago* Den. et Schiff.	36	4	BL	—
292	*Xanthia ocellaris* Bkh.	37	8	E	BL
293	*Aegle kaekeritziana* Hbn.	26	9	E	V
294	*Mesogona acetosellae* Den. et Schiff.	42	5	BL	BL
295	*Agrochola lychnidis* Den. et Schiff.	39	19	BL	BL
296	*Agrochola litura* L.	32	15	BL	BL
297	*Agrochola helvola* L.	41	4	E	—
298	*Agrochola lota* Clerck	36	9	E	BL
299	*Agrochola circellaris* Hfn.	37	5	E	—
300	*Agrochola humilis* Den. et Schiff.	38	6	BL	—
301	*Ammoconia caecimacula* Den. et Schiff.	42	10	BL	BL
302	*Conistra vaccinii* L.	32	16	E	BL
303	*Conistra rubiginosa* Scop.	35	6	E	—
304	*Conistra erythrocephala* Den. et Schiff.	38	7	E	—
305	*Eupsilia transversa* Hfn.	37	13	E	—
306	*Cosmia affinis* L.	31	7	BL	—
307	*Cosmia trapezina* L.	29	13	E	BL
308	*Cosmia pyralina* Den. et Schiff.	31	4	E	—
309	*Atethmia centrago* Haw.	34	4	E	BL
310	*Drybotodes tenebrosa* Esp.	35	9	E	—
311	*Aporophyla lutulenta* Den. et Schiff.	40	7	E	—
312	*Orthosia incerta* Hfn.	37	11	BL	BL

No.	Scientific names of species	A	B	C	D
313	*Orthosia miniosa* Den. et Schiff.	33	8	BL	—
314	*Orthosia cerasi* Fabr.	37	9	BL	—
315	*Orthosia cruda* Den. et Schiff.	27	10	BL	—
316	*Orthosia populeti* Fabr.	37	4	E	—
317	*Orthosia gracilis* Den. et Schiff.	37	10	E	BL
318	*Orthosia opima* Hbn.	37	5	E	—
319	*Orthosia gothica* L.	32	11	E	—
320	*Anorthoa munda* Den. et Schiff.	41	9	BL	—
321	*Egira conspicillaris* L.	39	13	BL	BL
322	*Tholera cespitis* Den. et Schiff.	37	15	E	BL
323	*Tholera decimalis* Poda	38	21	E	BL
324	*Anarta trifolii* Hfn.	32	5	BL	BL
325	*Polia nebulosa* Hfn.	50	4	E	—
326	*Proxellus lepigone* Mschl.	28	20	E	BL
327	*Pachetra sagittigera* Hfn.	44	7	E	BL
328	*Lacanobia w-latinum* Hfn.	39	17	BL	BL
329	*Lacanobia thalassina* Hfn.	36	11	BL	BL
330	*Lacanobia suasa* Den. et Schiff.	34	22	E	BL
331	*Lacanobia oleracea* L.	34	22	BL	BL
332	*Sideritis albicolon* Hbn.	42	14	BL	E
333	*Sideritis reticulata* Goeze	34	9	E	E
334	*Melanchra persicariae* L.	38	4	BL	—
335	*Melanchra pisi* L.	34	10	E	BL
336	*Hada plebeja* L.	33	12	E	—
337	*Mamestra brassicae* L.	41	21	BL	BL
338	*Hecatera dysodea* Den. et Schiff.	33	9	BL	BL
339	*Harmodia bicruris* Hfn.	35	18	BL	BL
340	*Conisania luteago* Den. et Schiff.	38	20	E	E
341	*Hadena rivularis* Fabr.	28	11	BL	BL
342	*Hadula dianthi* Wagner	35	8	E	BL
343	*Harmodia perplexa* Den. et Schiff.	31	13	E	—
344	*Hyssia cavernosa* Ev.	31	10	E	—
345	*Mythimna turca* L.	41	8	BL	—
346	*Mythimna pudorina* Den. et Schiff.	36	4	E	—
347	*Mythimna pallens* L.	32	22	BL	BL
348	*Mythimna vitellina* Hbn.	39	10	BL	BL
349	*Mythimna ferrago* Fabr.	37	7	BL	BL
350	*Mythimna l-album* L.	32	21	BL	BL
351	*Leucania obsoleta* Hbn.	38	12	BL	BL
352	*Peridroma saucia* Hbn.	50	11	BL	BL

No.	Scientific names of species	A	B	C	D
353	*Euxoa obelisca* Tutt	37	11	BL	BL
354	*Euxoa temera* Hbn.	32	10	BL	BL
355	*Euxoa aquilina* Den. et Schiff.	35	10	E	—
356	*Agrotis cinerea* Den. et Schiff.	36	7	E	E
357	*Agrotis exclamationis* L.	35	22	BL	BL
358	*Agrotis segetum* Den. et Schiff.	33	22	BL	BL
359	*Agrotis vestigialis* Hfn.	32	4	E	—
360	*Agrotis ipsilon* Hfn.	42	22	BL	BL
361	*Agrotis crassa* Hbn.	44	18	BL	BL
362	*Axylia putris* L.	29	21	BL	BL
363	*Ochropleura plecta* L.	27	21	BL	BL
364	*Parexarnis fugax* Tr.	35	5	E	—
365	*Diarsia rubi* Vieweg	30	6	E	BL
366	*Cerastis rubricosa* Den. et Schiff.	35	9	E	—
367	*Noctua pronuba* L.	50	22	BL	BL
368	*Noctua fimbriata* Schreber	47	14	BL	BL
369	*Noctua comes* Hbn.	41	4	E	—
370	*Noctua janthina* Den. et Schiff.	35	6	BL	BL
371	*Spaelothis ravida* Den. et Schiff.	45	8	E	BL
372	*Xestia xanthographa* Den. et Schiff.	33	11	BL	BL
373	*Xestia c-nigrum* L.	38	22	BL	BL
374	*Xestia triangulum* Hfn.	41	15	BL	—
375	*Eugnorisma depuncta* L.	40	10	BL	—
Nolidae (Average of wingspan is 23.2 mm)					
376	*Meganola albula* Den. et Schiff.	21	5	E	—
377	*Nola aerugula* Hbn.	17	7	E	—
378	*Pseudoips prasinana* L.	36	15	BL	BL
379	*Nycteola asiatica* Kruilkovsky	23	15	BL	BL
380	*Earias clorana* L.	21	14	E	BL
381	*Earias vernana* Fabr.	21	11	E	BL

Notes *Macrolepidoptera species collected successfully by* V *Visible or* BL *black light traps,* E *equal* N *serial number,* A *Wingspan (mm),* B *Network: Number of trap pairs,* C *Network: More efficient light source,* D *Nagytétény: More efficient light source.*

Table 1.
Macrolepidoptera species collected successfully by Visible or BL light-traps.

It is most remarkable, however, that the number of species for which the results of the national light-trap network could not detect a significant difference between BL and N traps was much smaller at Nagytétény where the BL trap was most frequently chosen by insects (**Figures 4–8**). So provided the moths are free to choose between traps placed extremely close to each other, they will fly to the BL trap. If the visible and BL traps are very close to each other, even the small moths

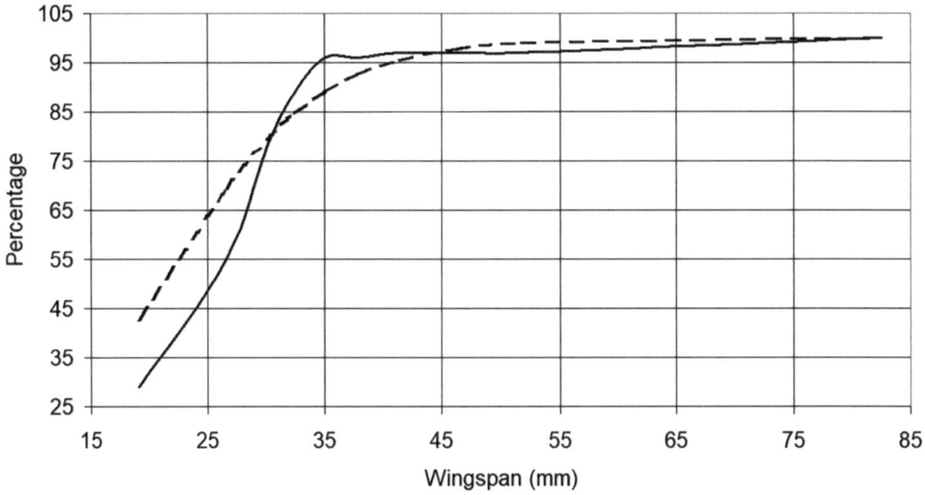

Figure 1.
Percentage of BL traps catch of macrolepidoptera species compared to the visible light ones in connection with the wingspan of moths (solid line = BL, dashed line = visible light).

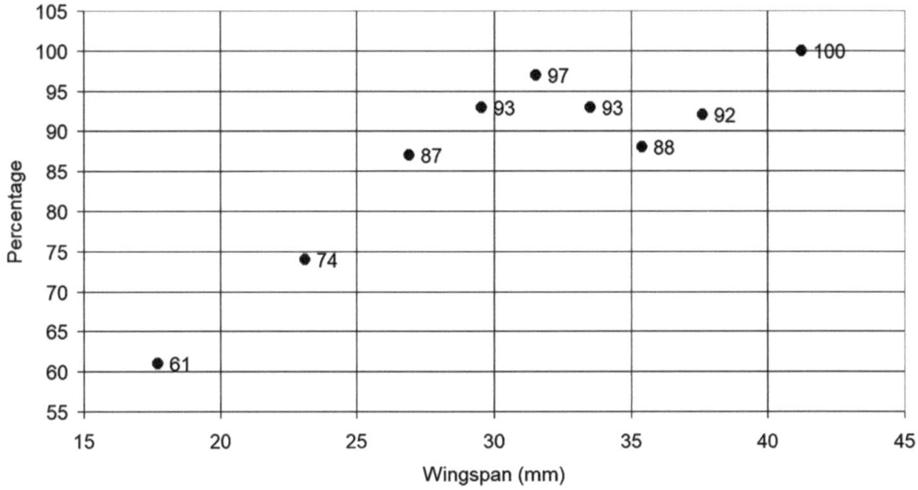

Figure 2.
Percentage of BL traps catch of macrolepidoptera species compared to the visible light ones in connection with the wingspan of moths (Nagytétény).

choose the BL traps en masse. However, such cases would be expected to be a random choice of the moths.

The fact that the highest number of moths with a wingspan greater than 35 mm, is in the BL traps, does not mean that these species cannot be collected with a visible bulb. However, it is clear that the visible or visible light source has low efficiency in collecting moths with wingspans greater than 35 mm. This result is noteworthy and can be used in plant protection and for another entomological research.

The light source of the trap should be chosen to suit our target species while bearing in mind their wingspan size.

The BL trap seems most efficient for operation for plant protecting purposes, despite the fact that their use is far more problematic.

Insect species are not only endangered by light trapping but also by the light pollution of urban areas. Our results confirm that the different light sources should

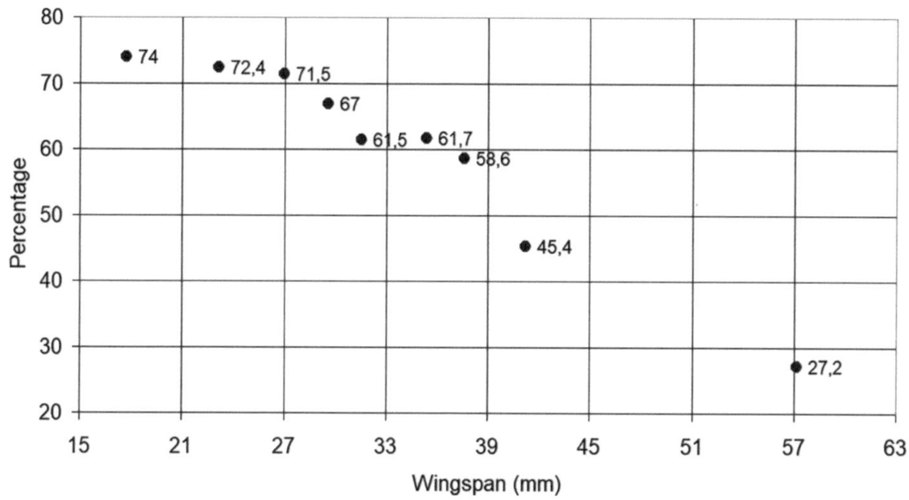

Figure 3.
Percentage of macrolepidoptera species caught by BL traps and visible light ones in connection with the wingspan of moths, if they select in equal proportion the two type light-traps.

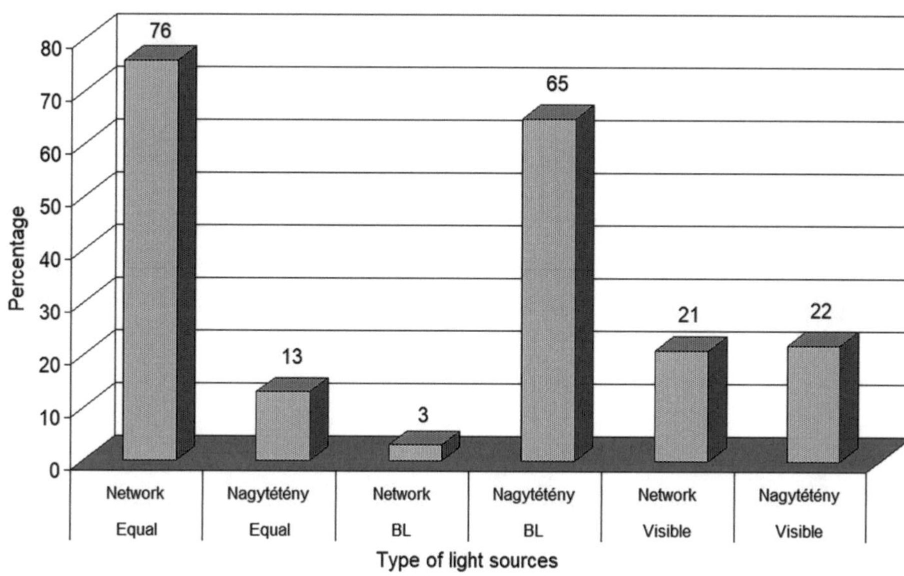

Figure 4.
Percentage of light-trap catch of Geometridae species by BL and visible light sources from data of the Hungarian light-trap network and Nagytétény.

incur mortality on different species to differing levels. Such differential mortality from artificial light sources could disturb the balance of life in biological communities. Kollings [52] established that there was a definite difference in the composition of the catch from two neighbouring street lamps. According to Frank [53], if some moth species are more attracted to light than others, the traits related to this attraction could help us to predict the effects of artificial light on communities of nocturnal species.

Light pollution might, in the future, expand to cover new areas. Some species may have populations more influenced by light pollution than others and some individuals might be more prone to it than others. This may generate a selective pressure to change behaviour. On the other hand, densely lit urban environments

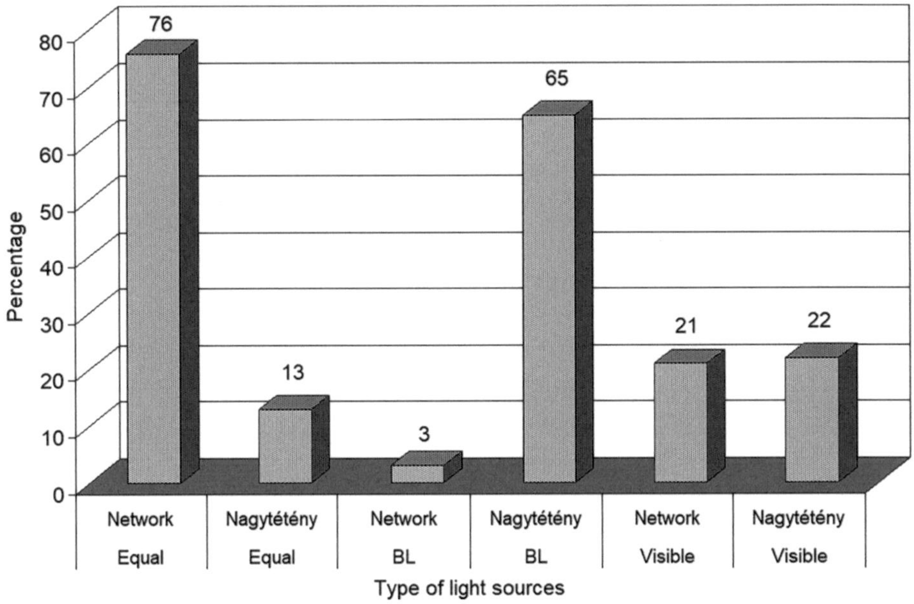

Figure 5.
Percentage of light-trap catch of Geometridae species by BL and visible light sources from data of the Hungarian light-trap network and Nagytétény.

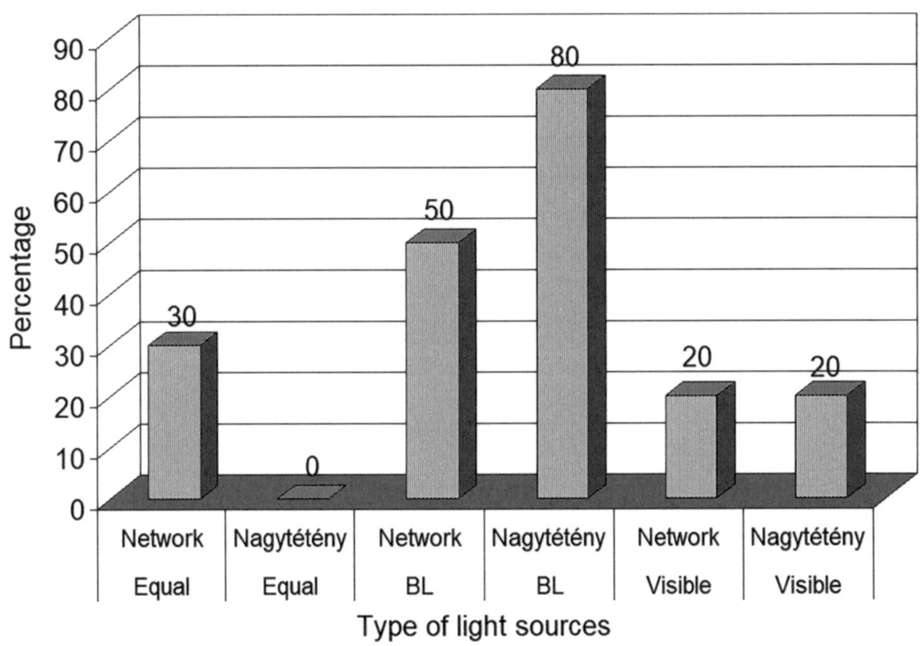

Figure 6.
Percentage of light-trap catch of Notodontidae species by BL and visible light sources from data of Hungarian light-trap network and Nagytétény.

may be advantageous for other species that fly by day or are not attracted to light. And there are also possibilities to solve the problem of light pollution. The use of low-pressure sodium lamps, for instance, may reduce the disturbing effects of illumination. These provoke a reaction of flying to light to a lesser extent than other

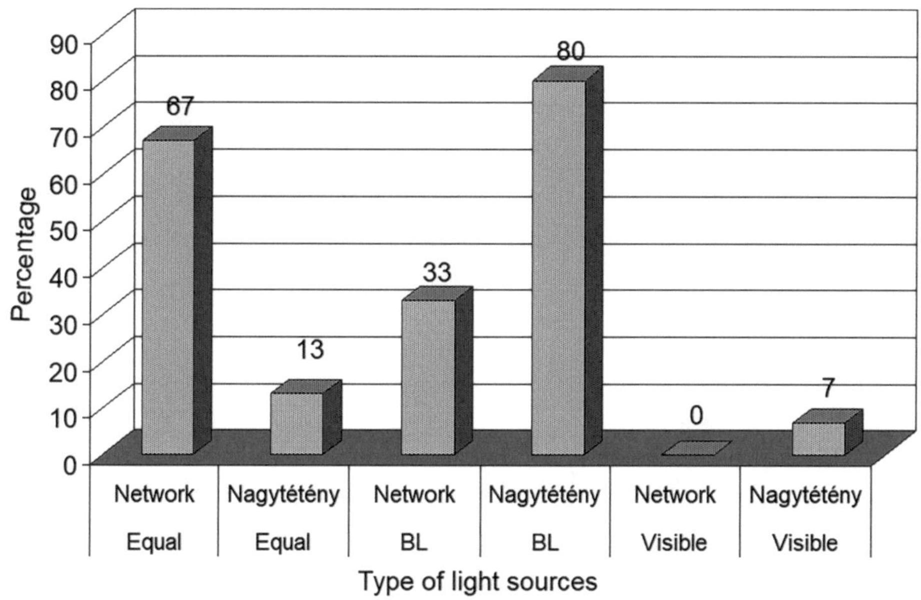

Figure 7.
Percentage of light-trap catch of Erebidae species by BL and visible light sources from the data of the Hungarian light-trap network and Nagytétény.

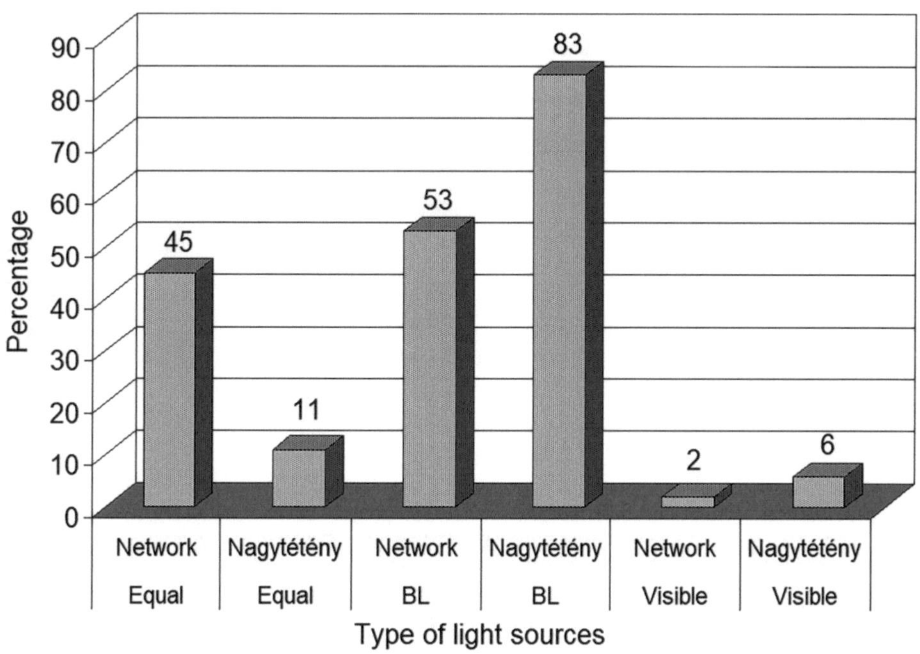

Figure 8.
Percentage of light-trap catch of Noctuidae species by BL and visible light sources from the data of the Hungarian light-trap network and Nagytétény.

lamps do. At the same time, they are also less likely to disturb the circadian rhythm of moths and other insects. These lamps also emit less energy than other lamps providing the same illumination. In an experiment by Eisenbeis and Hassel [54], the use of sodium vapour street lamps reduced the number of insects caught by 50%, including a 75% reduction in the number of moths.

Author details

László Nowinszky[1], Lionel Hill[2], János Puskás[1], Károly Tar[3] and Levente Hufnagel[4]*

1 Eötvös Loránd University, Savaria Campus, Savaria Science Centre, Budapest, Hungary

2 Biosecurity Tasmania, Hobart, Tasmania, Australia

3 University of Nyíregyháza Institute of Tourism and Science of Geography, Nyíregyháza, Hungary

4 John Wesley Theological College, Research Institute of Multidisciplinary Ecotheology, Budapest, Hungary

*Address all correspondence to: levente.hufnagel@gmail.com

References

[1] Mikkola K. Behavioural and electrophysiological responses of night-flying insects, especially Lepidoptera, to near-ultraviolet and visible light. Annales Zoologici Fennici. 1972;**9**: 225-254

[2] McFarlane JH, Eaton JL. Comparison of electroretinogram and electromyogram responses to radiant energy stimulation in the moth *Trichoplusia ni*. Journal of Insect Physiology. 1973;**19**(4):811-822

[3] Agee HR. Spectral sensitivity of the compound eyes of field-collected adult bollworms and tobacco budworms. Annals of the Entomological Society of America. 1973;**66**:613-615

[4] Pappas LG, Eaton JL. The internal ocellus of *Manduca sexta* and spectral sensitivity. Journal of Insect Physiology. 1977;**23**(1-2):1355-1358

[5] Eguchi E, Watanabe K, Hariyama T, Yamamoto K. A comparison of electrophysiologically determined spectral responses in 35 species of Lepidoptera. Journal of Insect Physiology. 1982;**28**(8):675-682

[6] Gui HL, Porter LC, Prideaux GF. Response of insects to color, intensity, and distribution of light. Agricultural Engineering. 1942;**23**:51-58

[7] Taylor JG, Deay HO. Electric lamps and traps in Corn Borer control. Agricultural Engineering. 1950;**31**(10): 513-505

[8] Frost SW. Response of insects to black and white light. Journal of Economic Entomology. 1954;**47**(2): 275-279

[9] Belton P, Kempster RH. Some factors affecting the catches of Lepidoptera in light traps. The Canadian Entomologist. 1963;**95**(2):832-837

[10] Sifter F. The use of the UV light trap for studying the flight of *Curculio* (*Balaninus*) *elephas* (in Hungarian). Növényvédelem. 1871;7(3):108-110

[11] Bürgés Gy, Nowinszky L, Puskás J, Herczig B. Comparative study of Macrolepidoptera caught in light-trap operating with normal and UV light. In: 55th International Symposium on Crop Protection. Ghent. 2003. p. 140

[12] Nabli H, Baily WC, Necibi S. Beneficial insect attraction to light traps with different wavelengths. Biological Control. 1999;**16**(2):185-188

[13] Mészáros Z. A comparison of the Microlepidoptera materials collected by normal and UV light-traps (in Hungarian). Folia Entomologica Hungarica. 1966;**19**(3):109-131

[14] Puskás J, Nowinszky L. Comparison of Macrolepidoptera data collected by normal and UV light-traps at Bakony-highlands (in Hungarian). Folia Musei Historico-Naturalis Bakonyiensis Zirc. 1994, 1994;**13**:89-105

[15] Day A, Reid WJ. Response of adult Southern Potato Wireworms to light traps. Journal of Economic Entomology. 1969;**62**(2):314-318

[16] Teel PD, VanCleave HW, Hollingsworth JP, Harstack AW. Spectral sensitivity of the Hickory Shuckworm to electromagnetic radiation. Journal of Economic Entomology. 1976;**69**(1):57-58

[17] Skuhravý V, Skuhravá M, Brewer W. The saddle gall midge *Haplodiposis marginata* (Diptera: Cecidomyidae) in Czech Republic and Slovak Republic from 1971-1989. Acta Societatis Zoologicae Bohemicae. 1993;**57**:117-137

[18] Cleve K. Einfluß der Wellenlänge des Lichtes auf den Lichtfang der

Schmetterlinge. In: Titschak E, editor. Deutscher Entomologentag in Hamburg. 1954;**30**(3):107–113

[19] Belton P, Kempster RH. Some factors affecting the catches of Lepidoptera in light traps. The Canadian Entomologist. 1963;**95**(2):832-883

[20] Jászainé VE. Miridae species (Heteroptera) in the material of normal and blacklight traps of Hungarian light-trap network in 1963 (in Hungarian). Folia Entomologica Hungarica. 1964;**17**: 471-524

[21] Theowald BR. Faunistische en fenologische waarnemingen met betrekking tot langpootmuggen (Diptera, Tipulidae). Fenologisch en faunistisch ondezoek over boom-gardinsekten. Wageningen. Versl. Landbounk. Onderz. 1963;**139**:185-202

[22] Jászainé VE. The swarming and distribution in Hungary of the cicadas spreading viruses *Laodelphax striatella* (Fallén) and *Javesella pellucida* (Fabricius, Homoptera, Areopidae) on the basis of the data of the national-wide light-trap network (in Hungarian). Növényvédelem. 1969;**5**(1):7-15

[23] Voigt E, Vojnits A. Observation of the species *Eupoecilia ambiguella* Hbn. and *Lobesia botrana* Den. et Schiff. by means of light-traps (in Hungarian). Növényvédelem. 1970;**6**(8):352-357

[24] Komlódi J. Biology of the Eurasian Hemp Moth (*Grapholita delineata* Walker) and results of control experiments (in Hungarian). Növényvédelem. 1970;**6**(8):343-348

[25] Blomberg O, Itämies J, Kuusela K. Insect catch in a blended and a black light-trap in northern Finland. Oikos. 1976;**27**:57-63

[26] Gál T, GY B, Eke I. Observation on the seasonal flight period of chestnut pests: a comparison of methods (in Hungarian). In: 23. Növényvédelmi Tudományos Napok. Budapest. 1976; **1976**:203-211

[27] Bürgés GY, Gál T. Zur Verbreitung und Lebensweise des Kastanienrüßlers (*Curculio elephas* Gyll., Col.: Curculionidae) in Ungarn. Zeitschrift für angewandte Entomologie. 1981;**91**(4): 375–382

[28] Bürgés GY. Light-trap catch depending on the power, colour and height of the light sources (in Hungarian). In: 4th Hungarian Ecological Congress, Pécs, 1997. p. 43

[29] Járfás J, Szabó E, Sohajda I. Evaluation of meteorological factors influencing the signalization of Silver Y (Autorgapha gamma L.) based on light-trap (in Hungarian) Kertészeti Egyetem Közleményei. 1975;**29**:167–174

[30] Járfás J, Tóth J. Forecast and protection of the damaging Melolontha species in vineyard (in Hungarian). Szőlőtermesztési Agrokémiai Tájékoztató, Kecskemét. 1977;**3**(1):2-7

[31] Járfás J., Szabó E., Ladics L. Prediction of Codling Moth (Cydia pomonella L.) and the investigation of climatic factors modifying its swarming (in Hungary). In: Scientific Session of the Performances of "János Lippay". 1977 p. 2396–2419

[32] Járfás J, Viola M. Experience gained in the light-trap observation of the Pea Podborer (Etiella zinckenella Treitschke) (in Hungarian). A Kertészeti Egyetem Közleményei. 1984; 221–227

[33] Járfás J, Viola M. The results of the observations of special light-traps for Beet Webworm (Loxostege sticticalis L.) (in Hungarian). Kertgazdaság. 1991;**4**: 65-70

[34] Járfás J. Results of light trapping of harmful tortricid moths (in Hungarian).

Kertészeti Egyetem Közleményei. 1977; **41**:123-126

[35] Jarfas J. The effectivity of different light trapping methods in studying the flight activity of European Corn Borer (in Hungarian). Növényvédelem. 1978; 494–498

[36] Wallner WE, Humble LM, Levin RE, Baranchikov YN, Cardé RT. Response of adult lymantriid moths to illumination devices in the Russian Far East. The Journal of Economic Entomology. 1995;**88**(2):337-342

[37] Fayle TM, Sharp RE, MEN M. The effect of moth trap type on catch size and composition in British Lepidoptera. The British Journal of Entomology and Natural History. 2007;**20**:221-232

[38] Barghini, A. Influência da Illuminãçao Artificial sobre a Vida Silvestre: técnicas para minimizar os impactos, com especial enfoque sobre os insectos. Universidade de São Paolo, Instituto de Biosciências, Programa de Pós-Graduãçao em Ecologia. 2008. p. 229

[39] Zollikofer CPE, Wehner R, Fukushi T. Optical scaling in conspecific *Cataglyphis* ants. The Journal of Experimental Biology. 1995;**198**: 1637-1646

[40] Jander U, Jander R. Allometry and resolution of bee eyes (Apoidea). Arthropod Structure & Development. 2002;**30**(3):179-193

[41] Spaethe J, Chittka L. Interindividual variation of eye optics and single object resolution in bumblebees. The Journal of Experimental Biology. 2003;**206**: 3447-3453

[42] Kapustjanskij A, Streinzer M, Paulus HF, Spaethe J. Bigger is better: implications of body size for flight ability under different light conditions and the evolution of alloethism in bumblebees. Functional Ecology. 2007; **21**:1130-1136

[43] Rutowski RL., Gislén L., Warrant E. J. Visual acuity and sensitivity increase allometrically with body size in butterflies. Arthropod Structure & Development 2009;**38**:91-100.

[44] Moser JC, Reeve JD, Bento JMS, Della Lucia TMC, Cameron RS, Heck NM. Eye size and behaviour of day- and night-flying leaf cutting and ablates. Journal of Zoology (London). 2004;**264**:69-75

[45] Yack JE, Johnson SE, Brown SG, Warrant EJ. The eyes of Macrosoma sp. (Lepidoptera: Hedyloidea): A nocturnal butterfly with superposition optics. Arthropod Structure & Development. 2007;**36**:11-22

[46] Kino T, Oshima S. Allergy to insects in Japan I. The reaginic sensitivity to moth and butterfly in patients with bronchial asthma. Journal of Allergy and Clinical Immunology. 1978;**61**(1): 10-16

[47] Barghini A, Medeiros BAS. Artificial Lighting as a vector attractant and cause of disease diffusion. Environmental Health Perspectives. 2010;**118**: 1503-1506

[48] van Langevelde F, Ettema JA, Donners M, MF WDV, Groenendijk D. Effect of spectral composition of artificial light on the attraction of moths. Biological Conservation. 2011; **144**(9):2274-2281. DOI: 10.1016/j.biocon.2011.06.004

[49] Hajtman B. Introduction to mathematical statistics for psychologists (in Hungarian). Akadémiai Kiadó, Budapest. 1968. p. 183

[50] Odor P, Iglói L. An introduction to the sport's biometry (in Hungarian). ÁISH Tudományos Tanácsának Kiadása. Budapest. 1987. p. 267

[51] Manczel J. Statistical methods in agriculture. Mezőgazdasági Kiadó. Budapest. 1983. p. 496

[52] Kollings D. Ecological effects of artificial light sources on nocturnally active insects, in particular on butterflies (Lepidoptera). Faunistisch-Oekologische Mitteilungen Supplement. 2000;**28**:1-136

[53] Frank KD. Impact of outdoor lighting on moths. An assessment. Journal of the Lepidopterists' Society. 1988;**42**(2):63-93

[54] Eisenbeis G, Hassel F. Attraction of nocturnal insects to street lights - a study of municipal lighting systems in a rural area of Rheinhessen (Germany). Natur und Landschaft. 2000;**75**(4): 145-156

Chapter 4

Artificial Light at Night and Breast Cancer

Redhwan Al-Naggar and Lutfi Al-Maktari

Abstract

An increased risk of breast cancer has been associated with light at night (LAN). Several risk factors have been identified that play a crucial role in causing most cancers such as the consumption of alcohol and smoking, and low fruit and vegetable intake. As a risk factor for cancer, environmental factors also play an essential role including indoor and outdoor air pollution. Light pollution has been found to be a risk factor for breast cancer in the form of artificial light at night, with melatonin being the mediator between environment and the epigenome. The risk of cancer in night shift workers can also amplify by artificial light at night. Light at night may also effects sleep disruptions and also considered as a risk factor of breast cancer.

Keywords: artificial, light, night, breast, cancer, melatonin

1. Introduction

1.1 Historical background

Biological adaptation to the sun light has advanced over billions of years. Historically, 200 years ago was the discovery of electric both lighting and power for the first time. In spite of the fact that this invention has been of incredible advantage for humankind, it has completely changed work environments, social, and home.

Light at night gave us the in-dependency from the normal daylight for living and to work longer hours and with a more income. The utilize of electric has not only been limited to night time but has also moreover made it conceivable to live and work independently of daylight during the day.

The contrast between artificial lighting and the sunlight. Sunlight is strong at all twelve visible wave-lengths which cresting within the yellow region. However, artificial light has either extreme characteristic wave-length crests like fluorescent or shows a monotonic increment in irradiance as wave-length protracts like radiant.

Night light sources have the ability to light the evening sky brighter than the new moon. Many voices has started to complain about increase the night light and its health effects. In spite of the fact that the global increasing of breast cancer rates is observed in industrialized countries [1, 2].

Over the past decade the speculation that night light inhibit melatonin production may act as a risk of breast cancer. This may due to that exposure to night light disturbs endogenous melatonin secretion. There is presently evidence that these changes have had great effect on health globally. Exposing to light after sunset is overriding the natural light–dark exposure pattern [3–5].

1.2 Electric lighting and human health

Whereas broad light night is needed for cutting edge society and business. There are a few regulations to overcome this modern phenomena. Therefore, it is crucial to appraise the adverse health effects of night light on the human health so that effective intercession can be formed to reduce harmful effects of exposure to night light.

1.3 Introduction to artificial light

Night light is a human-made light which can delivered from lighting that radiates from electric lamps. The increase of night light in cities has extended human human activity into the dark hours. It has been assessed that more than 80% of the world population is beneath light-polluted atmosphere that have an excessive artificial light [1, 6, 7].

Shifting to the light emitting diode (LED) technology in cities world wide has consequence in increasing in artificial light at night and especially blue-light because the use of white LED as the brand-new light standard in urban and rural areas [8]. Its well known that shift work affect circadian disruption is likely a carcinogenic risk factor to humans [9]. Most studies focused on breast cancer. However, there are studies focused on prostate cancer and night and showed that light night likely to increase risks for prostate cancer [10–12].

The possible explanation is that the night light inhibit the melatonin production which is a hormone usually produced at the dark, changes of sleep activities, and deregulation of circadian genes [13, 14]. Melatonin affects on estrogen-receptor positive human breast cancer cells [15–17].

In addition, several studies about day and night shift workers showed an interruption in both peak and lower melatonin levels in their urine compared to day workers. Furthermore, individuals reading using devices that produced blue light such as e-Book, surfing internet with their smart phones before going to bed compared with those who is reading a printed book will take longer to sleep [18].

Genetic background plays a role in changes in wake and sleep time [19]. For instance, a study showed that the lowest melatonin levels was reported among night-shift workers with the morning preference chronotype [i.e. The chronotype of a person is the individual's ability to sleep over a 24-hour cycle at a given time]. Other factors that may contributed are sex, age, living indoors, type of personality, nighttime illumination may also be related to chronotype [20].

2. Light pollution

The main achievement of the past century can be called artificial light. Nevertheless, while artificial lighting is extremely advantageous for modern society. It also poses a major drawbacks to human health and the ecological system of the environment in the form of pollution. It is possible to clearly describe light pollution as unnecessary and misdirected artificial lighting [21]. Light pollution defined as a situation where improper use of night light may hinder with people's comfort and health [22]. Other than the radical issue of sky glow caused by artificial lighting [23], a variety of known ecological problems are also caused by light pollution. Health problems in humans caused by exposure to night light are the most urgent among the known consequences. Via the eye, light is introduced to the human body. While the majority of the light mediates other biological processes in humans, such as light and dark cycles. The First Atlas of Artificial Night Sky Brightness, reported that about 99% of European and American populations, and about one-fifth of world is under polluted skies [24].

2.1 Indoor artificial light

Increase evening exposure to indoor short-wavelength light from LED lights, television, tablete, and smart-phons all these consider to disruptive to circadian function. Electricity system has been improved in the last century due to that the electricity became very cheap [25]. Moreover, artificial light today increased by the wide use of solar panels. This make the night light available in the remote areas. Thus, population exposure to night light in both indoors and out-doors has increased substantially [26]. Unlike the regular light which had fixed and known emission spectra.

2.2 Outdoor artificial light

The outdoor light is considered the modern world work and lives in. However, most of the people live now in dim light all the day without exploration to the sun light and undimmed light at night. Outdoor artificial lights includes lighting of: streets, advertising, architectural, security, domestic, vehicles, highways, railways, residential and commercial buildings, industries, shopping centers and sport facilities [27].

Light at night emitted from the above sources, is captured by the satellites through the sensors, circling around the Earth and transferral the captured information into the Defence Meteorological Satellite Program's database. This database incorporate yearly data after excluding moon and sun light, and other sources of light like fires and lightning. These images capture a small portion of the light arise from the earth's surface light, they symbolize the levels of night light at the ground level.

Outdoor light at night is considered as a permeant environmental threat by stressing the environment and human health [3, 28, 29]. Theories speculate that extreme exposure of night light is contributing to several health problems by interrupt the circadian rhythm [30].

2.3 Light and human physiology

The dark-and-light cycle provides the basic basis for life on our earth. Human adjustment to the solar cycle has consequence in fundamental molecular and genetic endogenous processes that correspond with an 24-hour period. The circadian genetic clock mechanism is closely involved in many cellular and organs function. While near 24-hours rhythms are produced spontaneously by the circadian system, human master clock must be readjust daily by the light–dark cycle to ensure proper temporal synchronization with the environment.

This everyday entertainment is mainly accomplished in humans by novel photoreceptors that project straight to the site of the circadian clock. The cycle production of a light-sensitive endogenous rhythm probably developed to permit for precise 24-hour control of activities and rest, and also to adapt to seasonal changes in night-length changes, while retaining the benefits of the underlying physiology that anticipates day and night.

3. Human physiology and light

Light is the most important trigger for human circadian rhythms to be controlled and is the key environmental time signal for the circadian clock. Light also triggers addition neuro-endocrine and neuro-behavioral responses, including restraint of melatonin production, directly alerting the brain, and improving alertness. The most popular biomarkers studied of the human physiological consequence to dark–light is melatonin. Melatonin is correlate with darkness and is produced at night only, irrespective of whether human is day or night active.

The synthesis and timing of the output melatonin includes a suprachiasmatic nucleus (SCN) afferent signal. Operation of this pathway, which happen in patients with damage of upper cervical spinal, wholly get rid of melatonin production. Some other circadian rhythms not rely on this pathway such as cortisol, body temperature, and sleep–wake cycles.

Increased in tumor development is related to light at night exposure in human [7, 10, 31, 32]. Several studies showed an increased incidence of breast cancer among those living in a heavy artificially lighted areas, due to the effects to circadian and thus suppression of melatonin. Animal study showed a higher tumor and more malignant tumors in female mice exposed to artificial light at night [33].

Three decades ago, a research showed that there was a greater risk of developing breast cancer in women with a history of night shift work. These results contributed to the consideration of light at night is a characteristics of developed nations [34].

3.1 Health effects of disrupted circadian rhythms

Epidemiological studies are a vital component part of the evidence base needed to determine night-time light exposure increase cancer risk. However, these studies are observational studies and cannot provide the mechanism.

4. Animal studies

Most animal experimental studies showed an accelerate onset of development of mammary tumors followed by continuous bright light at night compared to control animals maintained 12 hours of dark and 12 hours of light.

A dose-dependent suppression of melatonin levels is a consequence from exposure to bright, fluorescent night light. Exposure to the dimmest light intensity at night decreased the peak of circulating melatonin and thus enhancement of tumor growth rates and metabolic activity of linoleic acid.

A study showed that nocturnal melatonin inhibit the growth of both estrogen receptor negative and positive (ER- & ER+). The linoleic acid and poly-unsaturated fatty acid are essential for the alteration of (ER-) tumors. Therefore, and it can be used as a biomarker of cellular growth [35].

In rat hepatoma models, exposure to night light induce tumor growth has been replicated. The opposite is also true, the progressive restoration of circulating melatonin by decreasing exposure to night light is followed by a pronounce decrease in tumor growth.

Its worth mentioning that the growth of xenografts perfused with blood sample collected in the 12-hours dark time was markedly reduced. In addition, the addition of melatonin receptor antagonist blood obtained during darkness eliminated the blood's ability to suppress perfused tumor growth and metabolic activity.

Moreover, the addition of a melatonin receptor antagonist to the blood obtained during dark time eliminated the ability to suppress perfused tumors growth and metabolic activity. There is also Some evidence that circadian disruption by chronic phase advancement may increase cancer growth in laboratory animals.

Melatonin exhibits anti-proliferative and antioxidant properties modulates both cellular and humoral responses and control epigenetic responses and play a role in the apoptosis of cancer cells and in inhibition of tumor angiogenesis [36–40].

5. Human studies

Experimental evidence from animal studies have linked circadian rhythm disruption and circulating melatonin concentrations to disease progression. However, there is indirect human data and it focused on epidemiological research. These epidemiological research, however, have strengths and limitations, particularly in exposure evaluation and suitable comparison groups. Breast cancer has received the most research has been conducted on breast cancer. The idea was first articulated that the increase of night light may connected to high risk of breast cancer in the developed nations. Thus rising incidence and mortality rate in the developed nations.

Potential mechanisms involve suppression of melatonin and circadian gene activity. This hypothesis would predict that non-day shift work would increase the risk, blind women would be at lower risk, risk would be inversely correlated with recorded sleep length, and the amount of nighttime light in the population would co-distribute with the incidence of breast cancer.

Based on non-day shift occupations studies and breast cancer, it was concluded that shift work involving circadian disturbance is likely to be carcinogenic to humans. Several studies supporting this findings and the WHO has defined shift work as a potential carcinogen due to the chronobiological disturbance [9, 41].

Numerous research about the relationship between breast cancer and light night have been reported mixed findings [42–45]. Two studies found no association between night work shift and breast cancer [44, 45]. A significant increased risk was found a case control study in Norwegian people with a history of regularly working five nights [42, 43, 46].

Another study showed that a lower risk of breast cancer was associated with increased urinary excretion of melatonin [47]. Non-shift staff with a regular sleep and wake cycle often experience exposure to light at night got disruption of circadian rhythms [48].

After bedtime lights out, it is not yet clear if the ambient background light could affect the circadian system from weak sources in the bedroom or outside light. A brief exposure at these levels does not have a noticeable effect in a laboratory setting, but long-term chronic exposure may.

Four studies have identified an association with a risk of breast cancer of any components of night light level in the bedroom [49–51]. The elevated risk estimate was statistically significant in two of them [50, 51].

In addition to the limitation of recall error, there is also the potentially important limitation of recall bias as a case–control design. There is a possibility that even a very low of light night exposure may have an effect over a long period of time is significant. Few people in the western world sleep in a complete darkness. The circadian cycle can be disrupted by repeated awakenings with low light exposure in the bedroom but any associated health effects are unknown [52].

6. Light pollution and breast cancer

The most common cancer among women is breast cancer [53]. Breast cancer incidence rates is increasing rapidly worldwide. The breast cancer incidence increasing rapidly in Western nations such as: Europe, North America, Canada, Australia and New Zeland.

7. Night light and breast cancer

Excessive exposure to night light may raise the risk of breast cancer and there are several mechanisms. The retrieval of non-image-forming photo-receptors in the retina is transferred to the pineal gland as neural information where it suppressing the melatonin secretion. Then changing in the affinity of the estrogen receptor and an increased susceptibility to hormone-dependent cancers like breast cancer [26, 54].

The other possible mechanism is that human can suffer from circadian rhythms disruption by engaging in night time activities which activated by night light contributing to increased susceptibility to breast cancer [26, 55].

According to the third possible mechanism, light at night can function as a general stressor and endocrine disruptor, this is another possible mechanism, particularly when changes in night light intensity are suddenly and unexpected [26, 56]. It was also proposed that melatonin is a mediator between the epigenome and environment [28, 57].

A study found that women exposed to extensive night light are vulnerable to a 12% rise in breast cancer risk. The areas also with the highest night light had a higher rate of occurrence of breast cancer compared to the areas with the darkest environments [31].

In a global study showed that Artificial light at night is significantly correlated for breast cancer. Immediate practical steps should be taken to ban artificial lighting at night in the major cities around the world and also in homes as well [10].

A Nurses' Health Study reported a small increased breast cancer risk among pre-menopausal women associated with exposure to residential outdoor light at night [30].

A study concluded in Georgia showed that there is a positive associations between breast cancer risk and exposure to night light. This is consistency with the previous research which showed that there is a biological associations between night light exposure to and breast cancer risk. Studies have indicated that disruption of circadian cycle may result in breast tissue carcinogenic. About 30 to 50% higher risk of breast cancer in developed countries with the highest versus lowest light night levels [31, 58].

Blue light exposure before sleeping has been shown to inhibit the development of nocturnal melatonin, which may be associated with an increased risk of breast cancers [13, 59, 60].

All documented reported an environmental impacts of artificial night light, like melatonin production are due to the spectrum of the light [61]. These physiological changes may influence the circadian cycle, sleep timing, control of blood pressure, seasonal reproduction and the role of melatonin as an antioxidant with implications for the prevalence of some certain cancers [15, 62].

The results were not related to a specification lighting technology, but to artificial night light. Studies showed an association between patterns of night light and the incidence of breast cancer and obesity [10, 31, 58, 63].

Blask and colleagues famously showed that a key factor in the connection is melatonin, a hormone produced in nighttime darkness that promotes sleep [64]. They found that when the tumors were perfused with melatonin rich human blood collected during the night, the development of cancer slowed down.

Some studies have concentrated on the CLOCK gene. Stevens and his colleagues found that healthy women displayed lower CLOCK gene expression than women with breast cancer. Switching on or off of genes as a result of artificial light at night may play an important role [65].

7.1 Melatonin

Melatonin was first discovered in 1917, indicating that an extract of the bovine pineal gland given to frogs whitened their seed coat (McCord and Allen 1917). Melatonin is a pleiotropic neuro-hormone produced through the darkish section of the 24 hours cycle and secreted by the endocrine gland called pineal gland. Even when the period of light is short with low intensity, melatonin secretion remains to be inhibited by light [37, 64, 66].

About 40 years later chemical extracted from the pineal gland identify and called it melatonin [67]. Since its discovery, melatonin existence has been noted in all studied organisms, ranging from single-cellars to all plants and animals [68].

Various central and peripheral tissues, including the heart and arteries, breast, lung, small intestine, liver, kidney, adrenal gland, ovaries, gallbladder, prostate, and skin. Melatonin also serves as an anti-inflammatory and antioxidant that functions as a free radical scavenger during inflammations and injuries. The rhythmic release of melatonin from the pineal gland helps organize circadian rhythms and neuroendocrine processes by activating pairs of MT1 and MT2 G-protein receptors. Melatonin triggers fatigue, sleepiness and a diminution of sleep latency [69].

Furthermore, it has been demonstrated that inhibition of melatonin in humans by light depends on wavelength, with illumination of shorter wavelengths which is greener and more blue, being more effective at suppressing melatonin than yellow–redness of longer wavelengths [62]. Overall, exposure to artificial light by disrupting melatonin production, interferes with our temporal organization, possibly leading to cell dysfunction, and promoting the abnormal proliferation of altered cells.

Several research showed that melatonin plays a significant role in numerous functions of living human and can be an antioxidant and anti-oncogenic agent [70, 71].

The production and secretion of pineal melatonin increases after dusk and, in humans, it resumes between 2 am and 4 am, while after this period, its production slows down and stops about 3 h after sunrise [72]. As we additionally know today, light disturb melatonin creation and secretion.

8. Melatonin functions

The immune system exhibits rhythms and disturbance daily. Association between serum melatonin and phagocyte activity was documented. The presence of lymphocytes melatonin receptors is another significant point for connection between melatonin, and changes in the immune system [73].

Among its different capacities, melatonin moves the dark sign to all body cells and tissues keeping up the first essential capacity of a photoreceptor creation, recognizing photophase and scotophase, while the relations between them is likewise moved by melatonin subsequently flagging irregularity to the cells and tissues.

In young people melatonin is known to stifle the development and activation of the reproductive system, thus delaying sexual maturity. Melatonin is also known as a direct anti-oncogenic agent In regards to breast cancer, several mechanisms are suggested for its role, including. Melatonin is also known as anti-oncogeneic agent for breast cancer and prostate cancer. Concerning breast cancer, modification of estrogen receptors is the key that melatonin play its role.

9. Melatonin and breast cancer

Clinical studies showed a relationship between melatonin levels and metastatic of breast cancer, where a decrease in melatonin were noted in breast cancer patients, when compared with healthy ladies, bigger tumors also related with lower melatonin levels.

Melatonin synthesis and release happens in response to darkness and inhibited by light. Melatonin peak levels normally occur during sleep after midnight. These levels decrease and become minimal when we are exposed to day light.

A call for additional investigation of the connection between night light and shift workers and cancer that might be melatonin depended. This call is as yet applicable for shift workers, urban and rural populations exposed to light at night, electronic gadgets with LED lights become a well known source of light pollution [36].

It has been showed that high melatonin levels at day time result in seasonal affective disorder which is very common in northern America and in northern Europe. This disorder is managed primarily mental doctors, and additionally by chrono-scholars, as a typical solution for Seasonal Affective Disorder is presenting the subjects to short frequency light which stifles melatonin. This demonstrating the significance of timing, the decrease of melatonin secretion at day time is also important for human well-being, while disturbance that may rise out of a deficient amount of light at day time will likewise have negative well-being impacts.

The importance of exposure to natural day light is of great importance for the resetting of melatonin production. The short light wavelength suppresses melatonin production and skin produces vitamin D. Because of the modern life, many individuals around the world are not exposed to natural light as they venture out form home for work before sunshine and come back after sunset. This way of life result in the way that melatonin production isn't totally smothered, suppressed, while vitamin D production is stifled. As a result of night light exposure, environmental temporal cues are not transferred correctly to the cells for adjusting our temporal organization.

Driven brightening in electronic gadgets encompasses us and goes with us even while sleeping. Therefore, children exposed to such brightening for a long time during the dark period during the time melatonin is to be created under dim conditions. As melatonin levels are not tested consistently, a large portion of people are unconscious of the risks involves.

Several studies suggested the existence of the "immune-pineal" axis. The pineal gland serves as an inflammation mediator to synthesis melatonin, where pro-inflammatory mediators inhibit the production of melatonin while anti-inflammatory mediate potentiate melatonin production [74–76].

Mechanisms of many diseases can be explaned by understanding the relationship between the pineal gland and immune response. In addition to being a photoperiodic information transducer, the pineal gland is also a constitutive participant in the intrinsic immune response [76].

10. Light, sleep and breast cancer

As a likely mechanism, sleep would possibly affects circulating hormones such as melatonin, cortisol, growth hormone, prolactin and insulin levels, which can be key factors in many disease processes, including breast cancer [77]. Thus, many studies have examined the association between sleep period and breast cancer incidence [78–81].

10.1 Circadian, sleep disruption and mammary oncogenesis

Circadian and sleep disruption might also additionally act on molecular, immunologic, cellular, neuroendocrine, and metabolic levels in methods that make contributes to development of breast cancer.

Mammary carcinogenesis stars with mutation of a healthy cell through alterations in immune and cellular processes, including:

1. Oxidative nuclear DNA damage.

2. Changes in cellular apoptosis.

3. Changes in innate immune function.

Mutated cells are capable to survive despite the innate tumor suppressor mechanisms through inactivation of the p53 gene which could reduced apoptosis of aberrant cells [82]. Steroid hormones play an important role in the growth of breast cells, and estrogen considered an oncogenic agent is well known. The availability and timing of melatonin is adversely affected by sleep disturbance and circadian disruption.

Continual inflammation can stimulate genetic instability that can result in cell mutations. Disruptions in sleep and circadian cycle have affect on predispose the organism and metabolic processes to increased adiposity, which increase the risk of breast cancer.

11. Inflammation and the immune system

Repeated sleep and circadian cycle have an impact on innate and acquired immune function. Several studies have shown that decreased circulating NK cell numbers and NK cytotoxicity are associated with a single night of sleep deprivation [83, 84].

Sleep disruption may also causes a shift in cytokine creation from Th-1 cytokines to Th-2 cytokines, including IL-10, that may permit tumor cells to escape from immune surveillance and this may increased breast cancer risk [85]. Healthy sleep is associated with pre-myeloid dendritic cells production, which produce IL-12 a cytokine related to the delay of tumor onset [86].

Chronic inflammation is implicated in the oncogenic pathway, and both circadian disruption and sleep disruption were confirmed to have pro-inflammatory effects [87]. Certain lymphocytes are able to produce melatonin and in turn, melatonin has been proven to moderate the function of many immune functions [83].

Repeated disruptions in melatonin synthesis as a consequences of nightly circadian and sleep disruption might also additionally have negative effects on healthy immune functioning [88].

Sleep disruption is associated with increased glucocorticoid production and several studies indicated that short sleep duration and sleep deprivation are both associated with altered cortisol profiles and raised evening cortisol levels [89]. Glucocorticoid input from the adrenal gland has suppressive impacts on the amplitude of circadian cycle within the pituitary gland and hypothalamus [90].

12. Oxidative stress and DNA damage

Oxidative stress is known to cause DNA damage and is related with increased risk of breast cancer. On the other hand, antioxidants decrease DNA damage. Oxidative stress-induced DNA damage may be an essential pathway by which sleep disturbance may have effects on breast than circadian disruption [91].

Sleep have antioxidant impacts, giving an opportunity for the body to expel oxidants produced during alertness [92].

Many studies showed that sleep disturbance may increases reactive oxygen species (ROS) [93, 94]. ROS can lead to DNA damage by binding to neighboring molecules with unpaired electrons, causing strand breaks and its formation link between circadian cycle and sleep hardship [95].

Sleep disturbance contribute to oncogenesis through damage cellular, and disrupting the circadian cycle. Oxidative stress play an important role as a mediator of cellular apoptosis [96]; sleep deprivation-induced oxidative stress can permit the mutated cells to elude apoptosis. Circadian cycle also involved in this pathway.

13. Melatonin and estrogen

It has been proposed that the circadian cycle formed as a result of counteracting oxidative damage caused by radiation. Melatonin may be a powerful free radical forager that impacts quinine reductases to decrease oxidative damage [97].

14. Melatoin suppression

Studies showed that animals whose melatonin generation was disrupted by pineal or SCN ablation were prone to increased rates of mammary carcinogenesis. In another study, showed that melatonin suppression is related with increased hyperplastic processes and tumorigenesis in the mammary gland [98–100].

Several hormones, such as estrogens, progesterone and prolactin influence the etiology of breast cancer. Melatonin has an inhibitory effect on the receptors of estrogen and can counteract the estrogen's tumor-promoting impact. In addition, melatonin can increase the cell cycle length and avoid progression. These actions oppose the estradiol pathways, which promote cell proliferation and progression of cells [101, 102].

Sleep also impacts hormone levels.

Partial sleep deprivation among healthy women results in increases in Luteinizing Hormone (LH) and estradiol and elevated levels of prolactin levels [103]. Besides, there's conversely association between sleep hardship with levels of pestradiol. These indicated that regular sleep disturbance and short sleep at night may be related with chitinous elevation of estrogen levels which considered a risk factor of breast cancer. Ladies sleeping more than nine hours decreased the breast cancer risk compared to ladies sleeping seven hours [104].

15. Conclusions

The natural 24-hour cycle of light and darkness make a difference to preserve exact arrangement of circadian biological cycle, the main activation of the central nervous system and several cellular and biological processes, and pineal gland release of melatonin. All future outdoor lighting must be of energy efficient designs to reduce light pollution. The disturbance of the sleep–wake cycle and melatonin suppression may be due to to more direct health effects of night light. Even low intensity nighttime light has the capability of hinder melatonin release. Melatonin suppresses tumor growth and serves as a circulating anticancer signal. Many epidemiological studies back the theory that night light increases breast cancer risk. The quality and length of sleep and darkness affect several biological processes that are currently under investigation. Further information is required to assess the role of sleep versus the period of darkness on cancer. Due to the nearly present exposure to night light at improper times relative to endogenous circadian cycle, there is an urgent need for multidisciplinary research to address the assocation between night light and cancer.

Conflict of interest

There is no conflict of interest.

Author details

Redhwan Al-Naggar[1*] and Lutfi Al-Maktari[2]

1 Community Medicine Department, Faculty of Medicine, Al-Hikma University, Yemen

2 Haematology Department, Faculty of Medicine and Health Science, University of Sana'a, Yemen

*Address all correspondence to: radhwan888@yahoo.com

References

[1] Falchi F, Cinzano P, Elvidge CD et al (2011). Limiting the impact of light pollution on human health, environment and stellar visibility. J Environ Manage 92(10):2714-2722.

[2] Francies FZ, Hull R, Khanyile R, et al. (2020). Breast cancer in low-middle income countries: abnormality in splicing and lack of targeted treatment options. Am J Cancer Res;10(5):1568-1591.

[3] Davies T, Smyth T (2018). Why artificial light at night should be a focus for global change research in the 21st century. Glob. Chang. Biol. , 24, 872-882.

[4] Dominoni D, Jensen J, de Jong M, et al (2019). Artificial light at night, in interaction with spring temperature, modulates timing of reproduction in a passerine bird. Ecological Applications:e02062.10.1002/eap.2062.

[5] Stevens R (1987). Electric-power use and breast cancer-a hypothesis. Am J Epidemiol, 125(4):556-561.

[6] Falchi F, Cinzano P, Duriscoe D et al (2016). The new World Atlas of artificial night sky brightness. Sci Adv 2(6):e1600377.

[7] Garcia-Saenz, Ariadna et al (2018). Evaluating the Association between Artificial Light-at-Night Exposure and Breast and Prostate Cancer Risk in Spain (MCC-Spain Study). Environmental health perspectives. Doi:10.1289/EHP1837.

[8] Kyba C, Kuester T, Sánchez de Miguel A et al (2017). Artificially lit surface of Earth at night increasing in radiance

[9] Straif K, Baan R, Grosse Y, et al. Carcinogenicity of shift-work, painting, and fire-fighting. Lancet Oncol 2007;8:1065e6.

[10] Al-Naggar R and Anil S (2016). Artificial Light at Night and Cancer: Global Study. Asian Pac J Cancer Prev, 17 (10), 4661-4664.

[11] Behrens T, Rabstein S, Wichert K et al (2017). Shift work and the incidence of prostate cancer: a 10-year follow-up of a German population-based cohort study. Scand J Work Environ Health 43(6):560-568.

[12] Papantoniou K, Castaño-Vinyals G, Espinosa A et al (2016). Breast cancer risk and night shift work in a case–control study in a Spanish population. Eur J Epidemiol 31(9):867-878.

[13] Chang AM, Aeschbach D, Duffy JF, et al (2014). Evening use of light-emitting eReaders negatively affects sleep, circadian timing, and next-morning alertness. Proc Natl Acad Sci USA 112(4):1232-1237.

[14] Escofet J and Bará S (2015). Reducing the circadian input from self-luminous devices using hardware filters and software applications. Lighting Res Technol 49(4):481-496.

[15] Korkmaz A, Reiter R (2008). Epigenetic regulation: a new research area for melatonin? J Pineal Res 44(1):41-44.

[16] Palmer A, Zortea M, Souza A, et al. (2020). Clinical impact of melatonin on breast cancer patients undergoing chemotherapy; effects on cognition, sleep and depressive symptoms: A randomized, double-blind, placebo-controlled trial. *PLoS One*. 2020;15(4):e0231379. Doi:10.1371/journal.pone.0231379.

[17] Hill SM, Belancio VP, Dauchy RT et al. (2015). Melatonin: an inhibitor of breast cancer. Endocr Relat Cancer 22(3):R183–R204.

[18] Rafique N, Al-Asoom L, Alsunni A, et al (2020). Effects of Mobile Use on Subjective Sleep Quality. Nat Sci Sleep;12:357-364. Doi:10.2147/NSS.S253375.

[19] Jones SE, Tyrrell J, Wood AR et al (2016). Genome– Wide Association Analyses in 128,266 Individuals Identifies New Morningness and Sleep Duration Loci. PLoS Genet 12(8):e1006125.

[20] Chellappa S, Morris C and Scheer F (2020). Circadian misalignment increases mood vulnerability in simulated shift work. Scientific reports; 10:18614.

[21] Khan Z, Yumnamcha T, Mondal G, et al. (2020). Artificial Light at Night (ALAN): A Potential Anthropogenic Component for the COVID-19 and HCoVs Outbreak. Front Endocrinol (Lausanne);11:622. Doi:10.3389/fendo.2020.00622.

[22] Seoul Metropolitan Government Enforcement Rule of the Ordinance on the Prevention of Light Pollution and Management of Formation of Good Light. 2018. Available online: http://legal.seoul.go.kr/legal/english/front/page/law.html?pAct=lawView&pPromNo=1442.

[23] Liu, M.; Zhang, B.G.; Li, W.S.; Guo, X.W.; Pan, X.H. Measurement and distribution of urban light pollution as day changes to night. Light. Res. Technol. 2017.

[24] Cinzano P, Falchi F, Elvidge C (2001). The first world atlas of the artificial night sky brightness. Mon Not R Astron Soc, 328(3):689-707.

[25] Ahuja D, Tatsutani M. (2009). Sustainable energy for developing countries. Available from: http://sapiens.revues.org/823.

[26] Haim A, Portnov B. (2013). Light pollution as a new risk factor for human breast and prostate cancers. Dordrecht: Springer. DOI: 10.1007/978-94-007-6220-6_9.

[27] European Commission (EC). (2014). Health effects of artificial light. Available from: http://ec.europa.eu/health/scientific_committees/opinions_layman/artificial-light/en/index.htm.

[28] Haim A, Zubidat A. (2015). Artificial light at night: melatonin as a mediator between the environment and epigenome. Phil Trans R Soc B 370:20140121.

[29] Cho, C.-H.; Lee, H.-J.; Yoon, H.-K.; Kang, S.-G.; Bok, K.-N.; Jung, K.-Y.; Kim, L.; Lee, E.-I. Exposure to dim artificial light at night increases REM sleep and awakenings in humans. Chrono Int. 2015, 33, 1-7.

[30] James P, Bertrand KA, Hart JE et al (2017). Outdoor Light at Night and Breast Cancer Incidence in the Nurses' Health Study II. Environ Health Perspect 125(8):087010.

[31] Bauer S, Wagner S, Burch J et al (2013). A casereferent study: light at night and breast cancer risk in Georgia. Int. J. Health Geog. 12: 23.

[32] Keshet-Sitton A, Or-Chen K, Huber E, Haim A. 2017. Illuminating a risk for breast cancer: a preliminary ecological study on the association between streetlight and breast cancer. Integr Cancer Ther 16(4):451-463

[33] Walker, W.H., II; Bumgarner, J.R.; Walton, J.C.; Liu, J.A.; Meléndez-Fernández, O.H.; Nelson, R.J.; DeVries, A.C. Light Pollution and Cancer. Int. J. Mol. Sci. 2020, 21, 9360. https://doi.org/10.3390/ijms21249360

[34] Stevens RG, Brainard GC, Blask DE, Lockley SW, Motta ME. Breast cancer and circadian disruption from electric lighting in the modern world. CA Cancer J Clin. 2014 May-Jun;64(3):207-18.

[35] Dauchy R, Xiang S, Mao L, et al. (2014). Circadian and melatonin disruption by exposure to light at night drives intrinsic resistance to tamoxifen therapy in breast cancer. *Cancer Res*. 2014;74(15):4099-4110. Doi:10.1158/0008-5472.CAN-13-3156.

[36] Brzezinski A. (1997). Melatonin in humans. N Engl J Med; 336:186-195.

[37] Ferlazzo N, Andolina G, Cannata A, et al. Is Melatonin the Cornucopia of the 21st Century?. Antioxidants (Basel). 2020;9(11):1088. Doi:10.3390/antiox9111088.

[38] Korkmaz A, Topal T, Tan D, et al (2009) Role of melatonin in metabolic regulation. Rev Endocr Metab Disord. 2009;10:261-270.

[39] Lissoni P, Rovelli F, Malugani F, et al. (2001). Antiangiogenic activity of melatonin in advanced cancer patients. Neuro Endocrinol Lett;22:45-47.

[40] Sainz R, Mayo J, Rodriguez C, et al. (2003). Melatonin and cell death: differential actions on apoptosis in normal and cancer cells. Cell Mol Life Sci;60:1407-1426.

[41] Straif K, Baan R, Grosse Y et al. (2007). Carcinogenicity of shift-work, painting, and fire-fighting. Lancet Oncol. 8: 1065-1066.

[42] Hansen J, Stevens R. (2011). Case–control study of shift-work and breast cancer risk in Danish nurses: Impact of shift systems. Eur J Cancer.

[43] Lie JAS, Kjuus H, Haugen A et al. (2011). Night work and breast cancer risk among Norwegian nurses: Assessment by different exposure metrics. Am J Epidemiol;173:1272-1279.

[44] Pesch B, Harth V, Rabstein S, et al. (2010). Night work and breast cancer - results from the German GENICA study. Scand J Work Environ Health;36:134-141.

[45] Pronk A, Ji B, Shu X, et al. (2010). Night-shift work and breast cancer risk in a cohort of Chinese women. Am J Epidemiol;171:953-959.

[46] Stevens R, Hansen J, Costa G, et al (2011). Considerations of circadian impact for defining 'shift work' in cancer studies: IARC Working Group Report. Occup Environ Med;68:154-162.

[47] Schernhammer ES, Hankinson SE (2009). Urinary melatonin levels and postmenopausal breast cancer risk in the nurses' health study cohort. Cancer Epidemiol Biomarkers Prev. 2009;18:74-79.

[48] Gooley J, Chamberlain K, Smith K, et al. (2011). Exposure to Room Light before Bedtime Suppresses Melatonin Onset and Shortens Melatonin Duration in Humans. Endocrinology;152:742.

[49] Davis S, Mirick D, Stevens R (2001). Night shift work, light at night, and risk of breast cancer. J Natl Cancer Inst;93:1557-1562.

[50] Kloog I, Portnov BA, Rennert HS, Haim A. Does the modern urbanized sleeping habitat pose a breast cancer risk? Chronobiol Int. 2011;28:76-80.

[51] O'Leary E, Schoenfeld E, Stevens R, et al. (2006). Electromagnetic Fields and Breast Cancer on Long Island Study Group. Shift work, light at night, and breast cancer on Long Island, New York. Am J Epidemiol;164:358-366.

[52] Hatonen T, Alila-Johansson A, Mustanoja S, et al. (1999) Suppression of melatonin by 2000- lux light in humans with closed eyelids. Biol Psychiatry;46:827-831.

[53] World Cancer Research Fund International (WCRFI). (2014). Breast cancer statistics Available from: http://

www.wcrf.org/cancer_statistics/data_specific_cancers/breast_cancer_statistics.php.

[54] Blask DE, et al (2011). Circadian regulation of molecular, dietary, and metabolic signaling mechanisms of human breast cancer growth by the nocturnal melatonin signal and the consequences of its disruption by light at night. J Pineal Res. 51:259-269.

[55] Navara KJ, Nelson RJ. (2007). The dark side of light at night: physiological, epidemiological, and ecological consequences. J Pineal Res. 43:215-224.

[56] Ashkenazi L, Haim A. (2012). Light interference as a possible stressor altering HSP70 and its gene expression levels in brain and hepatic tissues of golden spiny mice. J Exp Biol. 215: 4034-4040.

[57] Schwimmer H, Metzer A, Pilosof Y, et al (2014). Light at night and melatonin have opposite effects on breast cancer tumors in mice assessed by growth rates and global DNA methylation. Chronobiol Int. 31:44-150.

[58] Kloog I, Stevens R, Haim A, et al (2010). Nighttime light level co-distributes with breast cancer incidence worldwide. Cancer Cause Control. 21:2059-2068.

[59] Gringras P, Middleton B, Skene DJ et al (2015). Bigger, brighter, bluer-better? Current light-emitting devices – adverse sleep properties and preventative strategies. Front Public Health 3:233.

[60] Stevens R, Zhu Y (2015). Electric light, particularly at night, disrupts human circadian rhythmicity: is that a problem? Phil Trans R Soc Lond B Biol Sci 370(1667):20140120.

[61] Aubé M, Roby J, Kocifaj M (2013). Evaluating potential spectral impacts of various artificial lights on melatonin suppression, photosynthesis, and star visibility. PLoS One 8(7):e67798.

[62] Cajochen C, Münch M, Kobialka S et al (2005). High sensitivity of human melatonin, alertness, thermoregulation, and heart rate to short wavelength light. J Clin Endocrinol Metab 90(3):1311-1316.

[63] Wyse C, Selman, Page M, et al (2011). Circadian desynchrony and metabolic dysfunction; did light pollution make us fat? Medical Hyp. 77: 1139-1144.

[64] Blask DE, et al. (2005) Cancer Res 65(23):11174-11184; doi:0.1158/0008-5472.CAN-05-1945.

[65] Hoffman AE, et al. Cancer Res 70(4):1459-1468 (2010); doi:10.1158/0008-5472.CAN-09-379

[66] Brainard GC, Hanifin JP, Greeson JM, et al (2001). Action spectrum for melatonin regulation in humans: evidence for a novel circadian photoreceptor. J Neurosci 21(16):6405-6412.

[67] Lerner AR, Case JD, Takahashi Y (1958) Isolation of melatonin, a pineal factor lightens melanocytes. J Am Chem Soc 80:2057-2058

[68] Kohidai L, Vakkuri O, Keresztesi M, Leppaluoto J, Casaba G (2002) Melatonin in the unicellular Tetrahymena pyriformis: effects of different lighting conditions. Cell Biochem Func 20:269-272

[69] Tordjman S, Chokron S, Delorme R, et al. (2017). Melatonin: Pharmacology, Functions and Therapeutic Benefits. *Curr Neuropharmacol*;15(3):434-443. Doi:10.2174/1570159X14666161228122115.

[70] Poeggeler, B., Thuermann, S., Dose, A., Schoenke, M., Burkhardt, S., Hardeland, R., 2002. Melatonin's unique radical scavenging properties e roles of its functional substituents as revealed by a comparison with its structural analogs. J. Pineal Res. 33, 20e30.

[71] Tan, D. X., Reiter, R. J., Manchester, L. C., Yan, M. T., El-Sawi, M., Sainz, R. M.,... Hardeland, R. (2002). Chemical and physical properties and potential mechanisms: Melatonin as a broad spectrum antioxidant and free radical scavenger. Current Topics in Medicinal Chemistry, 2, 181-197

[72] Reiter R, Tan D, Fuentes-Broto L (2010). Melatonin: a multitasking molecule. Prog Brain Res;181:127-151.

[73] Córdoba-Moreno M, de Souza EDS, Quiles C, et al. (2020). Rhythmic expression of the melatonergic biosynthetic pathway and its differential modulation in vitro by LPS and IL10 in bone marrow and spleen. *Sci Rep*. 2020;10(1):4799. Doi:10.1038/s41598-020-61652-5.

[74] MacDonald I, Huang C, Liu S et al. (2020).Reconsidering the Role of Melatonin in Rheumatoid Arthritis. Int. J. Mol. Sci; 21, 2877; Doi:10.3390/ijms21082877.

[75] Markus RP, Ferreira ZS, Fernandes PACM, Cecon E. The immune-pineal axis: a shuttle between endocrine and paracrine melatonin sources. Neuroimmunomodulation. 2007; 14(3-4): 126.

[76] Markus, R. P., & Ferreira, Z. S. (2011). The immune-pineal axis: The role of pineal and extra-pineal melatonin in modulating inflammation. Advances in Neuroimmune Biology, 1(1), 95-104.

[77] Morris C, Aeschbach D, and Scheer F (2012). Circadian system, sleep and endocrinology. Molecular and Cellular Endocrinology, 349(1):91-104.

[78] Hurley S, Goldberg D, Bernstein L (2015). Sleep duration and cancer risk in women. Cancer Causes and Control, vol. 26(7): 1037-1045.

[79] Vogtmann E, Levitan E, Hale L et al. (2013). Association between sleep and breast cancer incidence among postmenopausal women in the Women's Health Initiative. Sleep, 36(10):1437-1444.

[80] Wang P, Ren F, Lin Y et al. (2015). Night-shift work, sleep duration, daytime napping, and breast cancer risk. Sleep Medicine, 16(4):462-468.

[81] Xiao Q, Signorello L, Brinton L, et al (2016). Sleep duration and breast cancer risk among black and white women. Sleep Medicine; 20:25-29.

[82] Raulet DH, Guerra N. Oncogenic stress sensed by the immune system: role of natural killer cell receptors Nat. Rev. Immunol. 2009; 9(August):568-580

[83] Ruiz F, Rosa D, Zimberg I, et al. (2020). Night shift work and immune response to the meningococcal conjugate vaccine in healthy workers: a proof of concept study. Sleep Med;75:263-275. Doi: 10.1016/j.sleep.2020.05.032.

[84] Born J, Lange T, Hansen K, Molle M, Fehm HL. Effects of sleep and circadian rhythm on human circulating immune cells J. Immunol. 1997; 158(May):4454-4464.

[85] Dimitrov S, Lange T, Tieken S, Fehm HL, Born J. Sleep associated regulation of T helper 1/T helper 2 cytokine balance in humans Brain Behav. Immun. 2004; 18(July):341-348.

[86] Dimitrov S, Lange T, Nohroudi K, Born J. Number and function of circulating human antigen presenting cells regulated by sleep Sleep. 2007; 30(April):401-411.

[87] Irwin MR, Olmstead R, Carroll JE. Sleep disturbance, sleep duration, and inflammation: a systematic review and meta-analysis of cohort studies and experimental sleep deprivation Biol. Psychiatry. 2015

[88] Cole SW, Sood AK. Molecular pathways: beta-adrenergic signaling in cancer Clin. Cancer Res. 2012;18(March):1201-1206.

[89] Miller MA, Kandala NB, Kivimaki M, Kumari M, Brunner EJ, Lowe GD. Gender differences in the cross-sectional relationships between sleep duration and markers of inflammation: Whitehall II study Sleep. 2009; 32(July):857-864.

[90] Koyanagi S, Suyama H, Kuramoto Y, Matsunaga N, Takane H, Soeda S. Glucocorticoid regulation of 24-hour oscillation in interferon receptor gene expression in mouse liver Endocrinology. 2006;147(November):5034-5040.

[91] Kang DH. Oxidative stress, DNA damage, and breast cancer AACN Clin. Issues. 2002; 13(November):540-549.

[92] Nakata A, Ikeda T, Takahashi M, Haratani T, Fujioka Y, Fukui S. Sleep-related risk of occupational injuries in Japanese small and medium-scale enterprises Ind. Health. 2005; 43(January):89-97.

[93] Ramanathan L, Gulyani S, Nienhuis R, Siegel JM. Sleep deprivation decreases superoxide dismutase activity in rat hippocampus and brainstem Neuroreport. 2002; 13(August):1387-1390

[94] Silva RH, Abilio VC, Takatsu AL, Kameda SR, Grassl C, Chehin AB. Role of hippocampal oxidative stress in memory deficits induced by sleep deprivation in mice Neuropharmacology. 2004; 46(May):895-903.

[95] Boonstra J, Post JA. Molecular events associated with reactive oxygen species and cell cycle progression in mammalian cells Gene. 2004; 337(August):1-13.

[96] Buttke TM, Sandstrom PA. Oxidative stress as a mediator of apoptosis Immunol. Today. 1994;15(January):7-10.

[97] Poeggeler B, Reiter RJ, Tan DX, Chen LD, Manchester LC. Melatonin, hydroxyl radical-mediated oxidative damage, and aging: a hypothesis J. Pineal Res. 1993; 14(May):151-168

[98] Ird EA. The effect of experimental hypothyroidism on the development of mastopathia and tumors of the mammary gland in rats Probl. Endokrinol. Gormonoter. 1966; 12(January):99-102.

[99] Anisimov VN. Blastomogenesis in rats with persistent estrus Vopr. Onkol. 1971; 17:67-75.

[100] Tamarkin L, Cohen M, Roselle D, Reichert C, Lippman M, Chabner B. Melatonin inhibition and pinealectomy enhancement of 7, 12-dimethylbenz(a) an-thracene-induced mammary tumors in the rat Cancer Res. 1981; 41(November):4432-4436

[101] Lippman M, Bolan G, Huff K. The effects of estrogens and antiestrogens on hormone-responsive human breast cancer in long-term tissue culture Cancer Res. 1976; 36(December):4595-4601.

[102] Cos S, Gonzalez A, Martinez-Campa C, Mediavilla MD, Alonso-Gonzalez C, Sanchez-Barcelo EJ. Estrogen-signaling pathway: a link between breast cancer and melatonin oncostatic actions Cancer Detect. Prev. 2006; 30:118-128

[103] von TK, Norman TR, Armstrong SM. Overnight human plasma melatonin, cortisol, prolactin, TSH, under conditions of normal sleep, sleep deprivation, and sleep recovery J. Pineal Res. 1996; 20(January):7-14.

[104] Wu AH, Wang R, Koh WP, Stanczyk FZ, Lee HP, Yu MC. Sleep duration, melatonin and breast cancer among Chinese women in Singapore. Carcinogenesis. 2008 Jun;29(6):1244-8.

Chapter 5

Light Pollution Observations in Indonesia

Agustinus Gunawan Admiranto, Rhorom Priyatikanto, Siti Maryam, Elyyani, Siti Kurniawati and Muhammad Faisal Eko Saputro

Abstract

Light pollution is a growing concern in the world. It affects many walk of lives, including human health, the degradation of nocturnal animal habitat, and the inability of the astronomers to observe dimmer objects. We in Space Science Center of LAPAN (Indonesian National Institute of Aeronautics and Space) try to mitigate this through a coordinated observation of light pollution using Sky Quality Meter equipments which are located in several LAPAN's stations [Agam (West Sumatra), Pontianak (West Kalimantan), Sumedang (West Java), Garut (West Java), Pasuruan (East Java), Kupang (East Nusa Tenggara), and Biak (Papua)]. The observations has been conducted since 2018 in stationary and moving modes, and the results are then sent to a central database which is located in Space Science Center in Bandung (West Java). The results showed that there are some variations of light pollution across Indonesia. In this respect most of the stations have moderate pollution as can be seen from the values of Biak, Agam, Sumedang, and Pontianak (20.0, 19.5, 19.6, and 17.7 mpsas respectively). On the other hand, the stations which are located near or in cities have high light pollution (Bandung and Pasuruan with 17.1 and 18.0 mpsas, respectively). A particular station (Garut) has low light pollution (20.6 mpsas). The data of these observations are presented in a website to be accessed by interested parties.

Keywords: light pollution, coordinated observations, Sky Quality Meter, moving observations, website

1. Introduction

Light pollution is a growing concern all over the world. It affects all ways of life, human and animals alike. People are more and more affected by light pollution, and some are getting health problems like lack of sleep, obesity, and breast cancer. Other problems are that people cannot see the beauty of night sky, the disruption of the circadian rhythm in human and animals, and the efficiency of energy in street lamps in cities.

Falchi et al. [1] analyzed light pollution all over the world. They made an atlas of light pollution in Europe, Asia, and North America. They observed that more than 80% of the world and more than 99% of the U.S. and European populations live under light-polluted skies. They concluded that this pose humanity with problems of unprecedented magnitude. Cinzano et al. [2] discussed the same problem and observed that this problem also affect many regions like Central Africa and Central Asia which was previously thought were free from this problem.

Light pollution is related with the sky quality, in which the better the sky quality means that the light pollution is the lesser. A good sky quality means that one can see much of the objects of the sky, including dimmer objects. The sky quality can be measured using Sky Quality Meter device which records the light emitted by the sky and calculates the incident radiation. Spoelstra [3] measured the sky quality of Arnhem, Holland using an SQM which equipped with a filter wheel. The observations conducted in blue, green, red, and infra red wavelengths. His results indicated that the level of light pollution in a certain location depends on the dominant wavelength of the sky brightness of the location.

In their effort to quantify light pollution, Cinzano and Falchi [4] reviewed indicators of light pollution and products available with recent observational and modeling techniques. They used LPTRAN (Light Pollution radiative TRANsfer) software package to predict the distribution of the night sky in any visible wavelength. This software can be applied in various atmospheric conditions and geographical differences of local conditions (cloud cover) and ground surfaces. The calculations were then applied to the cities of Verona, Vicenza, Padova and Venezia in Italy. They found that the condition of the atmosphere can contribute significantly to the light poluution and they are able to do this quantitavely.

Light pollution have been mapped based on satellite data. Cinzano et al. [5] created the first world map of artificial light of the world using data obtained by Defense Meteorological Satellite Program (DMSP). The conclusion of their analysis is that light pollution of the night sky appears to be a global-scale problem affecting nearly every country of the World. The problem is more severe in the United States, Europe and Japan, as expected. However the night sky appears more seriously endangered than commonly believed.

Light pollution had been observed locally also. Using operation line scan (OLS) data of DMSP satellite, Tavoosi et al. [6] analyzed the light pollution of Teheran in 1993, 1998, and 2003 and mapped the changes of light pollution in those years. They concluded that the changes of light pollution in Teheran did not evenly occurred and this information can be used to project the development of the city in the future. Bruehlmann [7] observed the light pollution in the city of Winterthur, Switzerland using a luxmeter. These observations were compared with the results of DSLR observations obtained in International Space Station, and it was concluded that the map obtained from lux meter observations is more accurate than obtained using DSLR observation, but there is a good correlation between this two observations, so DSLR observation can be use to determine local light pollution map if there are no observations using lux meter.

Analysis of the light pollution map was carried out by Butt [8] who examined light pollution maps in urban and rural areas of Pakistan using data captured by the Defense Meteorological Satellite Program (DMSP) satellite using the Operational Linescan System (OLS) equipment on the satellite. The areas covered by this study include several provinces in Pakistan, namely Punjab, Sindh, Khyber Pukhtoonkhwa, and Baluchistan. In this case, the data were collected twice a day in the period 2004–2009. This data is taken from an altitude of 830 km from the Earth's surface with a spatial resolution of 2.7 km. The data obtained by Butt [8] is then processed using Geographical Informational System (GIS) software. Light pollution parameters used are light that comes directly from urban areas and sky glow caused by these urban areas. Furthermore, other data used is cartographic data (road networks, topography, land surface coverage, and river networks). Using the data obtained over a period of 6 years (2004–2009) they can finally obtain a map of light pollution during that period. They concluded that there are some areas that experience changes in light pollution levels drastically, but there are also areas that have relatively little change in light pollution. Other researchers, namely Chalkias et al. [9] conducted an analysis of the observational data for the Athens and surrounding areas in Greece etc. during the period 1995–2000 using the same data used by Butt [8] above, namely using the DMSP satellite and OLS data contained in that satellite. They tried to calculate the change in light pollution intensity during the 5 years period. They concluded that there are areas around Athens that have experienced a drastic increase in light pollution, but there are also those that are not experiencing too much light pollution.

Albers and Doriscoe [10] made a model of light pollution and discuss its implications for the many dark sky parks in the United States. They made a map of the brightness of the sky in the zenith direction and based on the assumption that the brightness of the sky has a linear relationship with population density in a particular area and is inversely proportional to the power of 2.5.

This map is then compared with satellite data obtained from the Defense Meteorological Satellite Program (DMAP) satellite. Next they made use of the scale of the sky's brightness expressed in the limiting magnitude in the zenith direction proposed by Schaaf [11] and then compared the model they obtained with the observational data they collected from each location they chose.

The results of their comparison showed that the observational data matched what they got from the model, except for certain places such as very dark areas and brighter areas. In dark regions the observer cannot observe the weakest stars as predicted by the model, while for bright regions the observer can still observe stars that are weaker than predicted by the model.

Related to light pollution which occurred in a certain location, this depends on the process of light arriving to that location. Bara and Lima [12] conducted an analysis of the arrival of light pollution in one place, by calculating the weight of each light pollution emission from various light sources arriving at that point.

In this case they do the sum of the radiation emission that comes at one point from various sources by taking into account the absorption and refraction processes of various types of aerosols through which the light emission passes before reaching the specified point.

Here, the data used is data from the Suomi-NPP satellite which appears in the form of a map of the distribution of light irradiation in a particular area where the area being the subject is covered in that area. Furthermore, the area under review is converted into a light pollution transfer matrix and from this the contribution made by each matrix component is calculated at the point that is the subject of this analysis. In this case, Bara and Lima [12] take the case of the A Veiga area in the Galicia region, and then calculate the light pollution experienced by the Xares area which is part of the A Veiga region.

Furthermore, a more detailed satellite map is made of the light pollution received by each point from the A Veiga area. From the calculations they did for each point on the map they made, finally a map was created showing the light pollution coming from A Veiga that was experienced by each surrounding area. This analysis conducted by Bara and Lima [12] can be applied to the case described by Albers and Doriscoe [10] above to calculate how far a point experiences light pollution from surrounding points, and how far each point contributes on light pollution experienced as a whole.

Light pollution observations had also been conducted in Indonesia, but those mainly were conducted sporadically mainly in an island (Java) and not in an extended period of time [13–15]. In this respect, we need observations of light pollution which conducted in extended period of time and in an extended area. As an archipelagic nation which covers vast area of sea and land, Indonesia need night sky brightness observations which conducted in regions outside Java so the comprehensive map of light pollution in Indonesia can be obtained.

This article is a summary of works related to the night sky brightness observations in Indonesia after its installation in 2018. The purpose of this article is to present the current understanding of light pollution in Indonesia using observations in several locations in Indonesia. Preliminary reports and analyzes of these works can be seen at Admiranto et al. [16, 17]. Section 2 describes the equipment, data, and methodology of obtaining the data. Section 3 describes the data analysis, and results of this data analysis. Section 4 describes the web presentation of the observations, and Section 5 describes discussion and conclusions of this analysis.

2. Data and methodology

Indonesian National Institute of Aeronautics and Space (LAPAN) has several stations across Indonesia which are used to observe night sky brightness. These stations are: Agam (West Sumatra), Pontianak (West Kalimantan), Bandung (West Java), Sumedang (West Java), Garut (West Java), Pasuruan (East Java), and Biak (Papua) (**Figure 1**). Locations of these stations can be seen in **Table 1**.

In each station we put Sky Quality Meter manufactured by Unihedron Company to observe the night sky brightness. The Unihedron Sky Quality Meter are SQM-LU types which can make night sky observations continuously during the night. These equipments are of photodiode-based instruments which has effective field of view of ~20° and produce sky brightness data in mag/arcsec^2. All SQM record sky brightness magnitude in zenith direction from 5 pm to 7 am with sampling rate of one minute. **Figure 2** shows the equipment we use to record the sky quality.

We made observations of night sky brightness (NSB) continuously since 2018 (the observations were not started simultaneously for each station). These data were obtained under different sky conditions, during dark and full Moon, low and high cloud cover, and clear and overcast conditions. Because the NSB are different between moonlit sky and otherwise, we then classify the nights depend on the phases of the Moon. The data then sent to a server in Bandung and organized according to data sources (7 stations) through Virtual Private Network (VPN) to the server in Space Science Center in Bandung through this address https://bimasakti.sains.lapan.go.id/. The organization of this data transfer process can be seen in **Figure 3**.

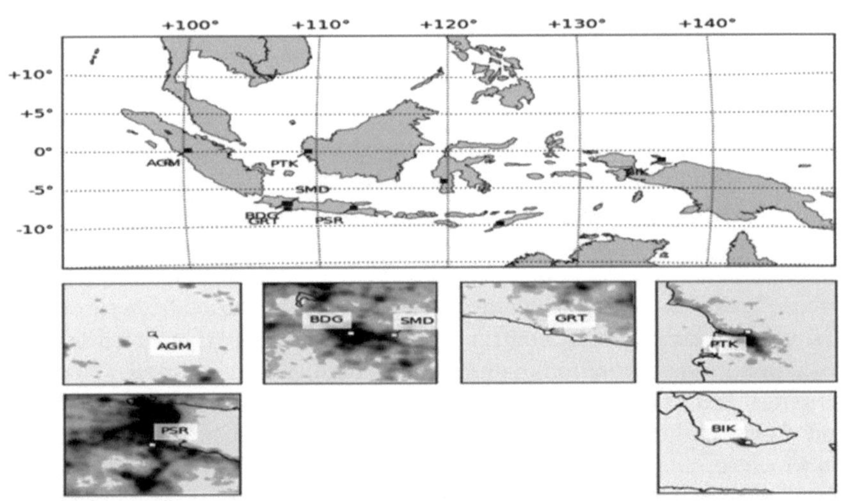

Figure 1.
Locations of sky quality observations in Indonesia.

Stations	Longitude	Latitude	Condition
Agam (AGM)	100 19′ 10.00″ E	0 13′ 48.00" S	Rural
Pontianak (PTK)	109 20′ 23.00″ E	0 02′ 48.00" N	Peri urban
Bandung (BDG)	107 40′ 40.21″ E	6 55′ 32.03" S	Urban
Sumedang (SMD)	107 50′ 13.97″ E	6 54′ 47.08" S	Rural
Garut (GRT)	107 41′ 31.97″ E	7 39′ 00.22" S	Rural
Pasuruan (PSR)	112 40′ 00.00″ E	7 34′ 00.00" S	Peri urban
Biak (BIK)	136 01′ 00.00″ E	1 17′ 00.00" S	Urban

Table 1.
Locations and conditions of 7 observation stations.

Figure 2.
Unihedron SQM LU-DL equipments.

The summary of the observations can be seen in **Tables 2–4**. The observations were not started simultaneously so there are some stations which were late in starting the observations. There are some occasions that the data was not obtained

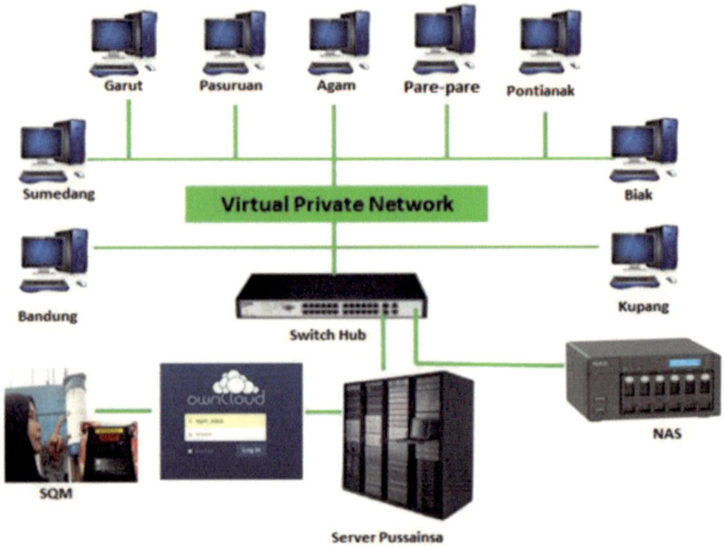

Figure 3.
Architecture of sky quality meter data management system in Space Science Center.

Location	2018											
	Jan	Feb	Mar	Apr	May	Jun	Jul	Aug	Sep	Oct	Nov	Dec
Agam	X	X	X	X	X	X	V	V	V	V	V	V
Biak	X	X	X	V	V	V	V	V	V	V	V	V
Kupang	X	X	X	X	X	X	X	X	X	X	X	V
Garut	V	V	X	V	V	V	V	X	V	V	V	V
Pasuruan	V	V	V	V	V	V	V	V	V	X	X	X
Pontianak	X	X	X	V	V	V	V	V	V	V	V	V
Sumedang	X	X	X	X	X	X	X	V	V	V	V	V

Table 2.
SQM data in 2018.

Location	2019											
	Jan	Feb	Mar	Apr	May	Jun	Jul	Aug	Sep	Oct	Nov	Dec
Agam	V	V	V	V	V	V	V	V	V	V	V	V
Biak	V	V	V	V	V	V	V	V	V	V	V	V
Kupang	V	V	V	V	V	V	V	V	V	V	V	V
Garut	V	V	V	V	V	V	V	V	V	V	V	V
Pasuruan	V	V	V	V	V	V	V	V	V	V	V	V
Pontianak	V	V	V	V	V	V	V	V	V	V	V	V
Sumedang	V	V	V	V	V	V	V	V	V	V	V	V

Table 3.
SQM data in 2019.

Location	2020											
	Jan	Feb	Mar	Apr	May	Jun	Jul	Aug	Sep	Oct	Nov	Dec
Agam	V	V	V	V	V	V	X	X	X	X	X	X
Biak	V	V	V	V	V	X	X	X	X	X	X	X
Kupang	V	V	V	V	V	X	X	X	X	X	X	X
Garut	V	V	V	V	V	X	X	X	X	X	X	X
Pasuruan	V	V	V	V	V	V	X	X	X	X	X	X
Pontianak	X	V	V	V	V	X	X	X	X	X	X	X
Sumedang	V	V	V	V	V	X	X	X	X	X	X	X

Note: V: available data, X: data is not available.

Table 4.
SQM data in 2020.

continuously, and this is caused sometimes by the equipment which was not working and we need to wait for the replacement from the vendor. Available data are designated by V, and unavailable data are designated by X.

Each station has its unique surroundings which make each has different light pollution characteristics. A station which is located near an urban concentration has high light pollution. We took satellite images of the stations from the site https://www.lightpollutionmap.info to verify the observations from below using the Sky Quality Meter. We can see that the sky conditions in Indonesia depend upon the proximities of those various locations to the source of night sky. There are places which are very dark, especially in Sumedang and Garut, but there are some places that are very bright like in Bandung and Pasuruan which are urban and peri urban in nature, respectively.

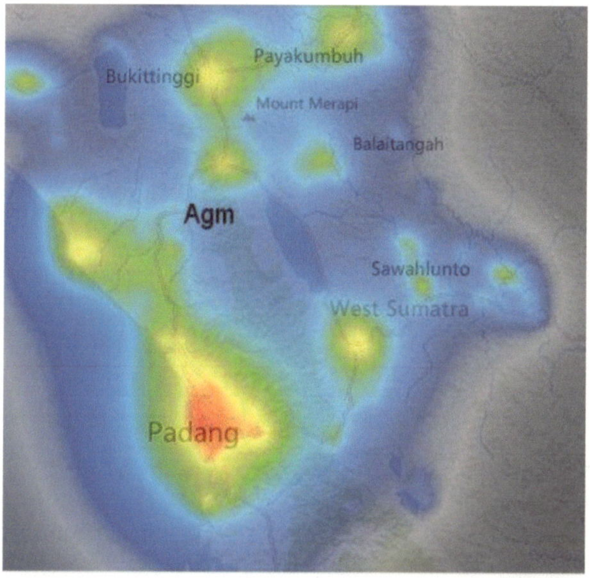

Figure 4.
Light pollution at Agam station.

2.1 Agam station

This station is located in some hilly region with height about 865 m above sea level, and the nearest city is about 20 km from the station. This explains the relatively low light pollution in this station at the value of 19.5 mag/arc sec^2. **Figure 4** is the light pollution of Agam and its surroundings. According to the satellite reading, the sky quality value observed from this station is 21.99 mag/arc sec^2.

2.2 Pontianak station

Pontianak station is located near the city of Pontianak (the capital of West Kalimantan), so the light pollution in this location is rather high as can be seen in the **Figure 5** below with the value of 17.7 mag/arc sec^2. The satellite reading of this station is 19.93 mag/arc sec^2.

2.3 Bandung station

Bandung is the capital of West Java at which is located about 750 m above sea level. so the light pollution here is a bit high as can be seen in **Figure 6** below with the value of 17.1 mag/arc sec^2. The satellite reading of this station is 19.55 mag/arc sec^2.

2.4 Sumedang station

Sumedang station which is about 800 m above sea level is a rural observation station, and the value of the sky quality is 19.6 mag/arc sec^2. Meanwhile, the satellite reading of this station is 21.00 mag/arc sec^2 (**Figure** 7).

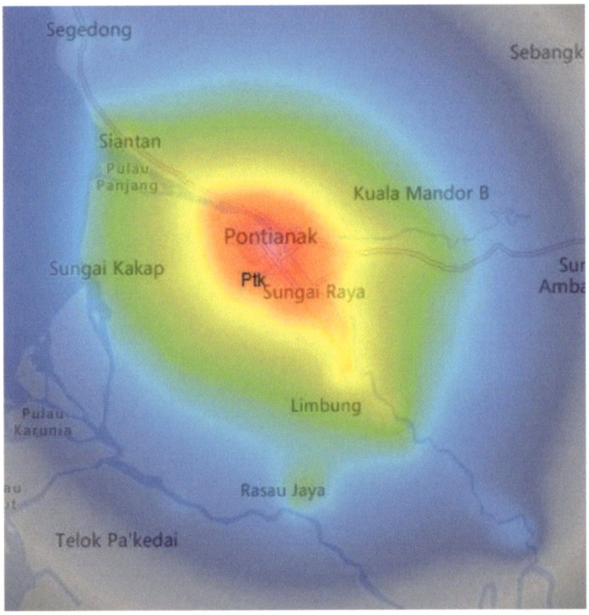

Figure 5.
Light pollution at Pontianak station.

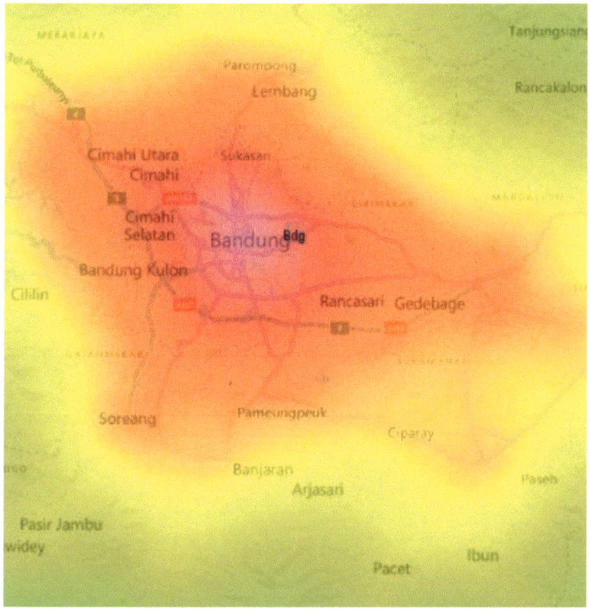

Figure 6.
Light pollution at Bandung station.

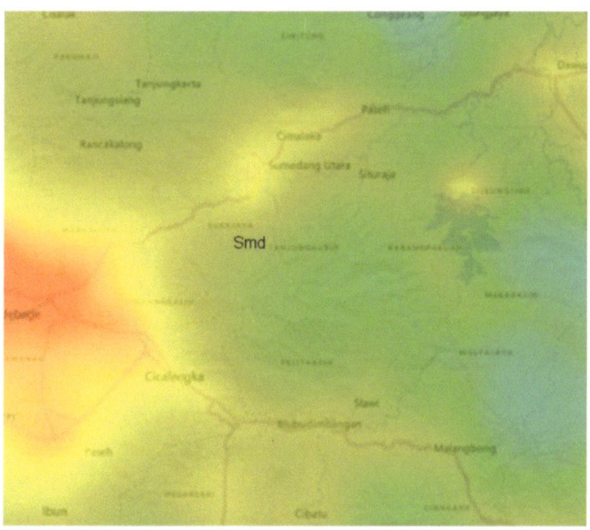

Figure 7.
Light pollution at Sumedang station.

2.5 Garut station

This station is located southern West Java coastal area. The value of the sky quality is 20.6 mag/arc sec^2, and the satellite reading of this station is 21.95 mag/arc sec^2 (**Figure 8**).

2.6 Pasuruan station

This station is located near a big city (Surabaya) and a busy highway. The value of the sky quality is 18 mag/arc sec^2, and the satellite reading of the sky quality is

Figure 8.
Light pollution at Garut station.

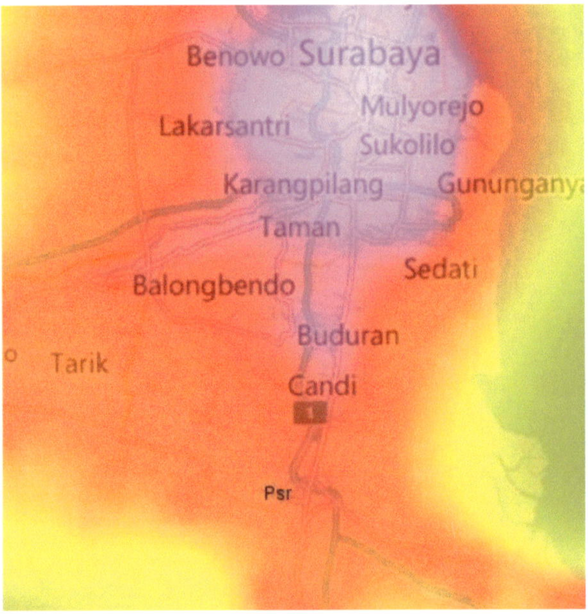

Figure 9.
Light pollution at Pasuruan station.

20.44 mag/arc sec^2. One can see in the **Figure 9** that the light pollution at this location is a bit high.

2.7 Biak station

This station is also located in a coastal area near the city of Biak, and the value of the sky quality is mag/arc sec^2. Meanwhile, the satellite reading of this station is 20.98 mag/arc sec^2 (**Figure 10**).

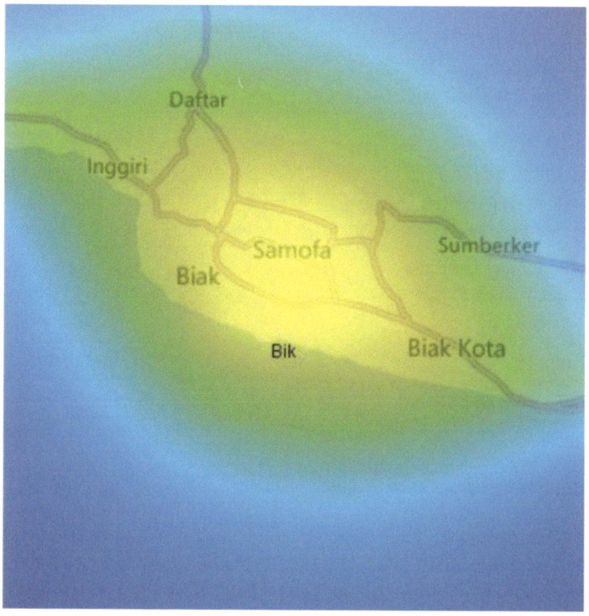

Figure 10.
Light pollution at Biak station.

S.No.	Station	S Q M (mag/arc sec^2)	Bortle Scale	Satellite (mag/arc sec^2)	Bortle Scale
1.	Agam	19.5	5	21.99	3
2.	Bandung	17.1	8	19.93	6
	Sumedang	19.6	5	21.00	4
	Garut	20.6	2	21.95	2
	Pontianak	17.7	8	19.93	6
	Pasuruan	18	6	20.44	5
	Biak	20	5	20.98	4

Table 5.
Comparison of sky quality observations using sky quality meter with reading from the satellite observations complemented with its Bortle scale.

Table 5 depicts the SQM observations and satellite reading of these stations. One can see that the value of the satellite reading of the sky quality is systematically higher than the value of the sky quality observed from below. Priyatikanto [18] stipulated that this discrepancies are caused by atmospheric extinction and scattering.

3. Moving observations

Apart of conducting stationary observations, we conduct observations in moving mode also. These mobile campaigns were conducted to measure the sky brightness using moving vehicles through various conditions, ranging from cities to villages, highlands to lowlands. The observations were conducted by six different team which observe the night sky brightness at the same night on June 24–26,

S.No.	Areas	Designation	Distance covered	Remarks
1.	Pasuruan-Probolinggo- Jember-Blitar -Mojokerto-Pasuruan	PSR	590 km	Mainly urban
2.	Surabaya-Lamongan-Bojonegoro-Ponorogo-Nganjuk-Pasuruan	SBY	532 km	Mainly urban
3.	Bandung-Cirebon-Semarang-Kudus- Ngawi -Salatiga-Kebumen	BDG	987 km	Mainly urban
4.	Agam-Bukittinggi-Batusangkar-Sawahlunto – Solok-Padang-Pariaman-Padang Panjang	AGM	292 km	Mainly rural
5.	Sumedang	SMD	94 km	Mainly urban
6.	Oelnasi-Kupang-Kuanheun	KOE	80 km	Mainly rural

Table 6.
Routes and distances covered in moving observations.

2019 during which the weather was relatively good (no rain or overcast) and the moonlight does not much contaminated the sky. During that period, the moon age was 20 to 23 days and the fraction of illumination was 63–45%.

In this campaign we use also the Unihedron SQM-LU (Sky Quality Meter with data logger attached through a USB port). The SQM was equipped with a battery charger and inserted into a tube that can be placed outside the vehicle. The Unihedron Device Manager application is also installed on a computer which connected to Shared GPS application on Android to see which location has such and such night sky brightness. Data were sampled all along the trip with 10 second intervals, and data recorded are location and time of observation, and sky brightness in unit of magnitude per square arcsecond (mag/arc sec^2), among others. The summary of observations are presented in **Table 6**.

In general, trips in **Table 6** which occurred in Java Island (1, 2, 3, and 5) were considered urban trips areas and the roads covered are in the northern part of Java which has more economic activities compared with the southern part of Java. On the other hand, trips in Sumatra and Timor (4 and 6) are considered mainly rural trips (**Figure 11**).

There are variations of night sky brightness among locations covered in the journey. The all journey which path of 2870 km length covered big cities with high sky brightness of 10 mag/arc sec^2 to the suburban and rural areas with low sky brightness of 21–22 mag/arc sec^2. The values of much of the data are in the 17–21 mag/arc sec^2 range.

Some of the data are contaminated by the glare of external light which made the sky brighter than it should be, but on the other hand some trees sometimes cover the equipment so the sky appear darker than it should be. This circumstances should be taken into account using some kind of data filtering to get the real sky brightness.

One of the purpose of moving observations of night sky brightness is to establish a preliminary sky brightness map. Night sky brightness data from VIIRS (Visible Infrared Imaging Radiometer Suite) satellite can be compared with SQM data to obtain an empirical model, in which the correlation can be used to obtain the night sky brightness over a greater area. **Figure 12** shows correlation between two kinds of data above. Density plots are shown to represent two dimensional data distribution in which darker color represent regions with higher density.

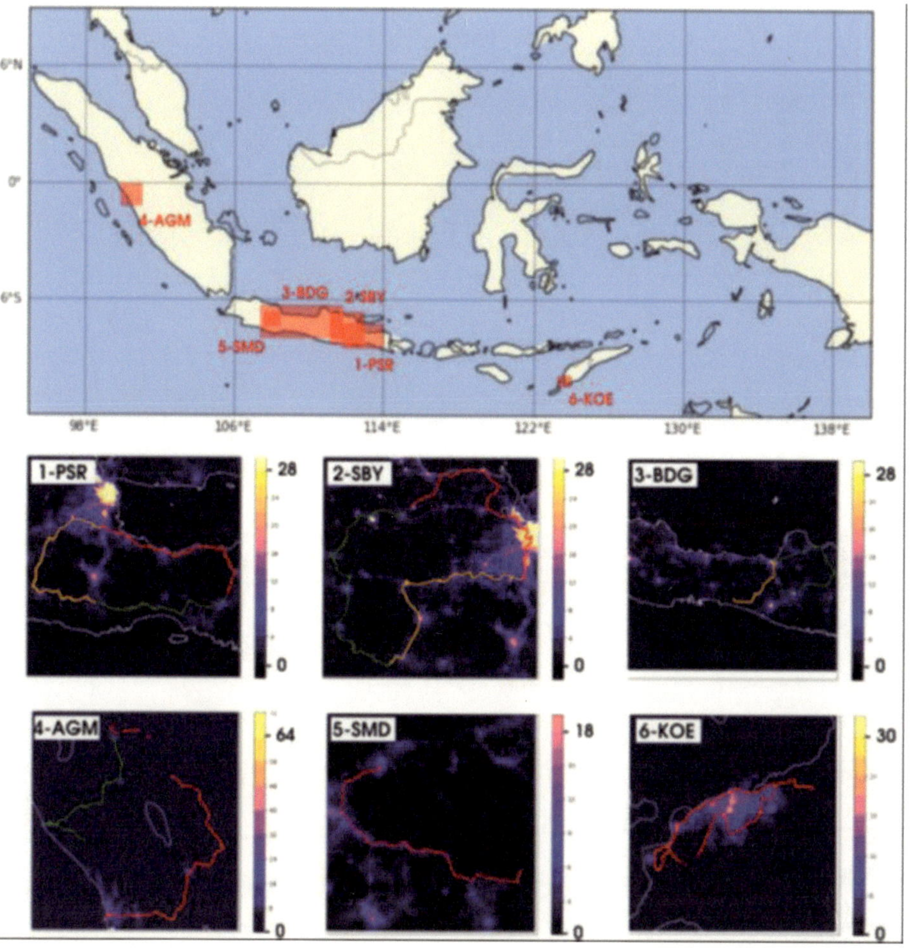

Figure 11.
Upper panel depicts regions covered in the campaigns. Lower panels show the routes taken by each team. Different colors (red, green, orange) represent different night of observation. This map is superposed with radiance map of VIIRS (in nW/m2/sr).

From the plots one can see a trend of increasing sky brightness with the increase of surface radiance measured by VIIRS. Nevertheless, this correlation is weak which caused by some contamination from unwanted light sources like light from transportation, street light, and buildings. Anoher problem is that the sky was not completely clear so some clouds scatter back the light some incoming radiation from the Earth's surface. We need very clear night condition to make observations with accurate results.

Priyatikanto et al. [19] obtained an empirical model in which the logarithm of VIIRS radiance and the value of night sky brightness in mag/arc sec^2 follow a linear function:

$$\mathrm{SQM} = 20.595 - 3.090 \log \mathrm{RADIANCE} \tag{1}$$

We can use the Eq. (1) to obtain the value of night sky brightness expressed in mag/arc sec^2 provided the value of VIIRS magnitude which should be expressed logarithmically.

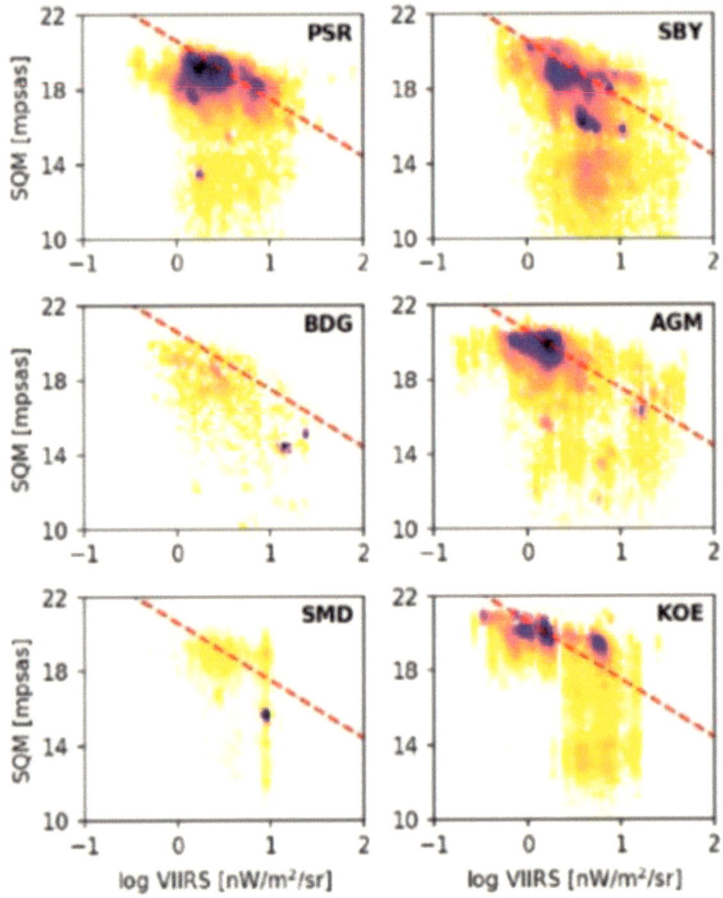

Figure 12.
Density plot of SQM data obtained during the trip by each team correlated with map from VIIRS which put in logarithmic scale. Dashed line represents empirical model from [14].

4. Website presentation

To disseminate the results of night sky brightness measurements, Space Science Center maintains a Sky Quality Observation (SQO) website which can be accessed at the address *sqm.sains.lapan.go.id*. This website is aimed for the general public where the information displayed uses simple language so that it can be easily understood even though they do not have an astronomical background.

The information displayed on the SQO Website includes understanding to readers regarding the explanation of the brightness of the night sky and light pollution. This information about the introduction of the brightness of the night sky is expected to arouse public awareness to protect the night sky in order to minimize light pollution. In addition, there is information to determine the brightness of the night sky using the Bortle Scale. The Bortle scale is a numerical scale to measure the brightness of the night sky. This scale starts from class 1 which is the darkest sky scale to class 9 which is the brightest sky which usually found in urban areas. This website also displays a glimpse of information about the Java-Bali light pollution map made based on the 2013 DMSP-OLS composite image combined with measurements of the Earth's surface as well as several location points for measurements of the brightness of the night sky carried out by the Space Science Center in Indonesia (**Figure 13**).

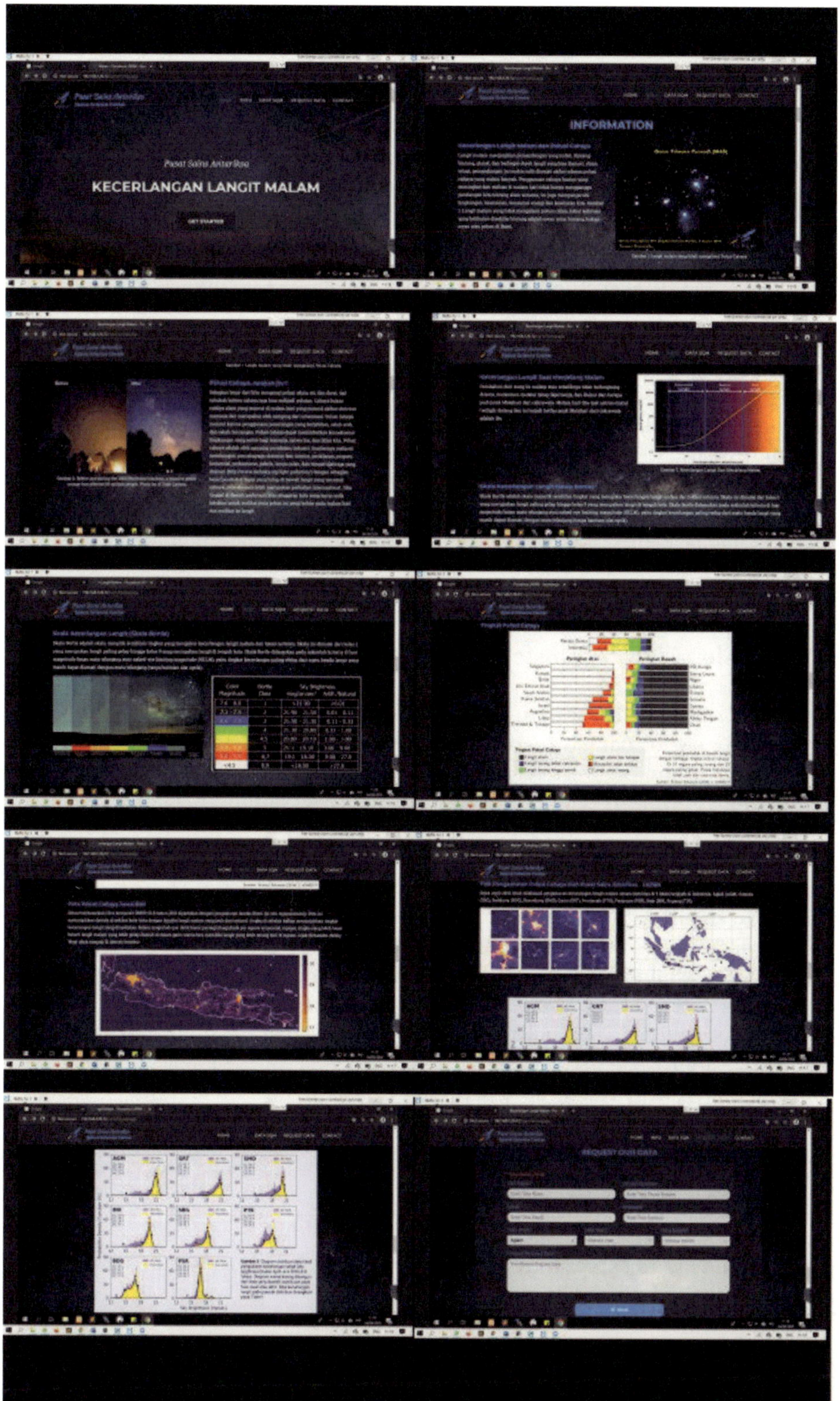

Figure 13.
Website representation about information of sky quality and its pollutions and of sky quality observations conducted by Space Science Center and LAPAN's stations.

There is a Request Our Data feature on the SQO Website, to facilitate ones which need night sky brightness data which had been carried out by the Space Science Center and the LAPAN's stations. To make a data request, one needs to fill personal information such as full name, telephone number, email address, and affiliation. After that, fill in the detail requested SQO data such as where the data from (Agam, Garut, Kupang, Pasuruan, Pontianak, and Sumedang) and the month-year of the data. And finally, fill in the reason for requesting SQO data.

Apart from providing information about the brightness of the night sky, the SQM website also has an SQO Map feature. This feature is presented digitally to make it easier for users to access by selecting data for the desired month and year. The information displayed in this feature is based on the average measurement of the SQO value each month, which is differentiated when calculating the average for the new moon and at the time of the full moon at points according to the location of the night sky brightness recorder and the point of observation location. Users can also zoom and scroll on the digital map to see the location of the night sky's brightness more easily.

Lastly, the SQO Website has a Backend or Admin Menu page that is used to manage content on the Front end of the Website such as Data Requests, Contact Pages, Homepage Info, Space and Atmospheric Observation Center locations, and SQO maps.

5. Discussions and conclusion

The coordinated observations of light pollution in Indonesia reveal some interesting insights, namely there are some regions in Indonesia with very low light pollution, and there are some sites which have very high light pollution. These observations are linearly correlated with satellite observations, namely which use VIIRS data channels.

The light pollution observations conducted by Space Science Center of LAPAN is the first coordinated light pollution observations in Indonesia. There are many works that should be done to improve the results, especially in the process of building of light pollution map across Indonesia. This can be done by installing more Sky Quality Meter equipments, especially near the locations with high pace of development to mitigate the adverse effects of light pollution.

On the other hand, light pollution observations should be conducted also in strategic locations, especially in some dark places which are suitable for astrotourism, and this strategy can be used to preserve the environment and can be used to make star party events to give some education in astronomy.

Acknowledgements

We would like to express our gratitude to Space Science Center of LAPAN to make the light pollution observations possible.

Author details

Agustinus Gunawan Admiranto*, Rhorom Priyatikanto, Siti Maryam, Elyyani, Siti Kurniawati and Muhammad Faisal Eko Saputro
Space Science Center, Indonesian National Institute of Aeronautics and Space (LAPAN), Bandung, Indonesia

*Address all correspondence to: gunawan.admiranto@lapan.go.id

References

[1] Falchi, F., Cinzano, P., Duriscoe, D., Kyba, C.C.M., Elvidge, C.D., Baugh, K., Portnov, B.A., Rybnikova, N.A., Furgoni, R. Science Advances; 2002 Vol 2, No. 6

[2] Cinzano, P., Falchi, F. Elvidge, C. D. Mon. Not. R. Astron. Soc. 2001; 328: 689-707

[3] Spoelstra, H. Journal of Quantitative Spectroscopy & Radiative Transfer. 2014; 139:82-89

[4] Cinzano, P., Falchi, F. Journal of Quantitative Spectroscopy & Radiative Transfer. 2014; 139:13-20

[5] P. Cinzano, P F. Falchi and C. D. Elvidge. 2001; Mon. Not. R. Astron. Soc. 328: 689-707

[6] Tavoosi, H., Darvishzadeh, R., Shakiba, A., Mirbagheri, B. Modelling Light Pollution in Suburbs Using Remote Sensing and GIS. In: The seventh International Conference on Urban Climate, 29 June - 3 July 2009, Yokohama, Japan; 2009

[7] Bruehlmann, S. M. Measuring and Mapping Light Pollution at a Local Scale, [dissertation], 2014; Vrije Universiteit Amsterdam.

[8] Butt, M. J. (2012), Estimation of Light Pollution Using Satellite Remote Sensing and Geographic Information System Techniques, GIScience & Remote Sensing, 2012, **49**, No. 4, p. 609-621. http://dx.doi.org/10.2747/1548-1603.49.4.609

[9] Chalkias, C., Petrakis, M., Psiloglou, B., Lianou, M. (2006), Modelling of light pollution in suburban areas using remotely sensed imagery and GIS, Journal of Environmental Management 79, 57-63

[10] Albers, S., dan D., Doriscoe (2001), Modeling Light Pollution from Population Data and Implications for National Park Service Lands, The George Wright Forum, Volume 18 No. 4

[11] Schaaf, F. (1994) The Schaaf Scale. *Sky and Telescope* (November), 64-65

[12] Bará, S., Lima, R. C. (2018) Photons without borders: quantifying light pollution transfer between territories, International Journal of Sustainable Lighting IJSL, 51-61

[13] Herdiwijaya, D. Light pollution at Bosscha Observatory, Indonesia. IOP Conf. Series: Journal of Physics: Conf. Series 1153. 2019; 012133 IOP Publishing doi:10.1088/1742-6596/1153/1/012133

[14] Pribadi, P., Pramudya, Y., Muchlas, Okimustava. The IoT implementation on the night sky brightness measurement in Banjar using the sky quality meter. AIP Conference Proceedings. 2202, 020023. 2019; doi:10.1063/1.51416362202

[15] Pramudya, Y., Budi, K. S., Okimustava, Muchlas. Preliminary study on relation between temperature, humidity and Night Sky Brightness in Yogyakarta. IOP Conf. Series: Journal of Physics: Conf. Series. 1231. 012004. 2019; IOP Publishing; doi:10.1088/1742-6596/1231/1/012004

[16] A. G. Admiranto, SK Haida, R Priyatikanto, S. Maryam, Ellyyani, N Suryana. "Mobile Campaign of Sky Brightness Measurement in Indonesia", Journal of Physics: Conference Series, 2020

[17] A. G. Admiranto, R. Priyatikanto, S. Maryam, Ellyyani, N. Suryana. "Preliminary Report of Light Pollution in Indonesia Based on Sky Quality Observation", Journal of Physics: Conference Series, 2019

[18] Priyatikanto, R Extinction and scattering parameters derived from sky

brightness measurements; Journal of Physics: Conference Series 1523 012017.2020; IOP Publishing; doi:10.1088/1742-6596/1523/1/012017

[19] Priyatikanto, R., Admiranto, AG, Putri, GP, Elyyani, Maryam, S and Suryana, N, Map of Sky Brightness over Greater Bandung and the Prospect of Astrotourism; Indonesian Journal of Geography, 51, 190. 2019; DOI: 10.22146/ijg.43410.

Chapter 6

Understanding the Importance of Harmonized Landscape Design for Forest Education Centers in Malaysia

Mohd Kher Bin Hussein and Nor Syuriaty Bt Jaafar

Abstract

The Forest Education Centre is a place that provides environmental education facilities where the community, teachers and students can visit to learn more about the forest environment. Therefore, landscape design elements' in this area which include buildings, should be visually in harmony with the sense of place and its surroundings. The design must concern form, color, materials, landscape degradation and preservation of natural processes. Unfortunately, landscape design elements in this area have been built based on in-situ approach where the absence of a 'code of design' in creating a well-built environment of landscape and architectural quality has contributed to inharmonious designs. This had led to the significant function of identity and meaning in securing the sense of place which was not adequately examined in the process. Therefore, the aim of this paper is to raise awareness among the stakeholders and the sensitivity towards the importance of harmonizing design in Forest Education Centres. Among the objectives of this study is to firstly understand the user's perception towards harmonies designs in a forest environment. Secondly, it is to suggest harmonies design approaches towards forest sustainability. The intended audience for this paper include directors, facilities managers, landscape architects, forest officers, architects and engineers.

Keywords: Forest, harmonize landscape design, education, sustainability

1. Introduction

Malaysia's forests are rich in species amidst an extremely complex ecosystem [1]. The world has focused on Malaysia's forests due to their functions, uniqueness and is considered as one of the world's mega-diverse countries ranked 12th in the world on the National Biodiversity Index [2]. Major forest types in Malaysia include the lowland dipterocarp forest, hill dipterocarp forest, upper hill dipterocarp forest, oak-laurel forest, montane ericaceous forest, peat swamp forest and mangrove forest. In addition, there are also smaller areas of freshwater swamp forest, heath forest, forest on limestone and forest on quartz ridges. Generally, forests play an important role in human life that makes it vital to be protected from deforestation and other negative impacts. Any unsustainable human interventions in a forest environment could also affect the climate, ecology, biodiversity, economy and health which in turn, will affect human life as well as other living things. Forests play a role in terms of climate when it puts up a defense against climate change and removes the greenhouse carbon dioxide gas. This helps to generate oxygen that assists in purifying the atmosphere and controlling rising temperatures. From the ecology aspect, forests prevent erosion by reducing the rainfall's force on the soil's surface, absorbing water and not allowing it to directly flow away and removing topsoil. Moreover, forests are storehouses of the planet's biodiversity, which includes flora and fauna, relating to food chains which are vital for human lives. Deforestation will destroy the food chains, thus, for economic benefits, the forest is a place for locals to generate their incomes through forest products trade such as forest fruits and fuel wood. Forests also provide opportunities for people to do outdoor or recreational activities such as walking, exercising, and jogging which would help them develop a healthier lifestyle.

The related regulations that affect forestry for Malaysia include the Land Conservation Act 1960, Environmental Quality Act 1974, National Parks Act 1980, Protection of Wildlife Act 1972, National Land Code 1965, Aboriginal Peoples Act 1954, Occupational Safety and Health Act 1994 and Forest Rules 1985 for Peninsular Malaysia. Meanwhile for Sabah, the relevant regulations include Forest Rules 1969, Wildlife Conservation Enactment 1977, Land Ordinance 1930, Cultural Heritage (Conservation) 1997, Sabah Parks Enactment 1984, Biodiversity Enactment 2000, Conservation of Environment Enactment 1996, Water Resource Enactment 1998, and Environmental Quality Act 1974. In addition, Sarawak has the Natural Resources and Environment Ordinance 1997, Forest Rules 1962, Wildlife Protection Ordinance and Rules 1998, The Forests (Planted Forest) Rules 1997, Sarawak Biodiversity Centre Ordinance 1997, Sarawak Biodiversity (Access, Collection & Research Regulations) 1998, Land Code 1958, Natural Resource and Environmental Ordinance, Water Ordinance 1994, Occupational Safety and Health Act 1994, Land Ordinance 1952, Native Code 1992, Native Code Rules 1996, and Native Custom Declaration 1996.

Furthermore, to ensure the forest areas in Peninsular Malaysia are protected, the Malaysian Government has classified the Permanent Reserve Forest (PRF) into various functional classes. Those classifications consist of Timber Production Forest, Soil Protection Forest, Soil Reclamation Forest, Flood Control Forest, Water Catchment Forest, Forest Sanctuary for Wildlife, Virgin Jungle Reserved Forest, Amenity Forest, Education Forest, Research Forest, Forest for Federal purposes, and State Parks [3]. The main aim of the classification is to manage the forest for various forest services and economic importance, while the forest quality is maintained to ensure the resources are sustainably managed. However, this would depend on the government's efforts in conserving and protecting the forests, which is not enough. We need to educate people about this matter and help all ages to understand and appreciate our country's natural resources and teach them how to conserve those resources for future generations. Therefore, the establishment of Forest Education Centers should be given credit since these centers enables people to realize how natural resources and ecosystems affect each other and how resources can be used wisely. Nevertheless, the development of these centers must be in harmony with the surrounding landscape.

Consequently, any development planned in the forest should consider the design aspect accordingly. The designs must be in harmony with the natural forest landscapes along with their own unique features that highlights the forest identity, besides being sites of recreational activities [4]. Furthermore, its designs should have good combinations between the natural environment and architectural aspects, for example, application of vernacular architecture [5]. Design of architectural structures (e.g. walkways, benches, signage's, shelters) and buildings, must be visually in harmony with the surrounding natural environment [6]. Forests with high panoramic values should have buildings with good style that is in harmony and does not contradict the existing environment [7]. Ceballos-Lascurain [8] and Walter [9] suggested that the most important consideration was that the structure and building designs in the forest must highlight aspects of the surroundings.

No doubt, the process of a new building or infrastructure at the Forest Education Center must be planned well before implementation to ensure natural landscape resources (e.g. Forests, rivers, hills, waterfalls, geological, etc.) are continuously protected and preserved. It is also to guarrantee that human made landscape elements can be in harmony with the existing environment. Buildings or infrastructure must conform to environment-friendly, low-impact architecture; renewable including solar energy, waste recycling, rainwater harvesting, natural cross-ventilation, no use of asbestos, controlled sewage disposal, and merging with the surrounding landscape. The design should enhance local architecture, use local materials as well as portray the local art and culture.

This was the main concern with the Forest Education Center as it is in their discussions to develop landscape designs, especially the landscape architectural images. The aim of this paper is to raise awareness among the stakeholders and the sensitivity towards the importance of harmonizing design in Forest Education Centres. Among the objectives of this study is to firstly understand user's perception towards harmonious designs within the forest environment. Secondly, it is to suggest harmonies design approaches on Forest Education Centres in Malaysia towards forest sustainability.

2. Forest education centre

The forest education centre is a place which provides environmental education facilities that community, teachers and students can visit to learn more about the forest environment, its importance and what people can do to help protect it. This centre promotes conservation of natural ecosystems, education focused on forest stewardship, and research related to forest ecology and management. Therefore, development of this site should be carefully focused on restoration, preservation, conservation, protection of natural systems and maintaining nature for its sustainability. Bear in mind that a sustainable forest environment is a forest that is carefully focused on restoration, preservation, conservation, protection of natural systems, maintaining natural function and the structures/buildings harmonizes with the landscape. Meanwhile, landscape design elements' developments such as buildings, shelters, benches, signage, information boards, roads, bridges, walkways, picnic tables and stairs should portray their identity and is in harmony with the sense of place. The design should have a good combination of a natural environment and architectural characteristics, such as the application of vernacular architecture. The designs should be visually in harmony with the surrounding natural environment and concerns form, color, materials, landscape degradation, preservation of natural processes, and protection of biological diversity.

One of the latest forest education centres in Malaysia is located in Ayer Hitam Forest Reserve, Puchong, Selangor. This area is also known as the Sultan Idris Shah Forestry Education Centre (SISFEC) and was awarded to Universiti Putra Malaysia through a long-term agreement to conduct activities regarding education, research and development in the field of forestry. SISFEC covers an area of about 1, 1761.1 hectares and has become an educational reference centre, research and development centre for good practice in tropical forest management at national level, as well as globally. This centre acts as an outdoor laboratory for students in their efforts to develop skills in forest management where they can gain more knowledge and skills in classifying trees and plants, learning about soil science, tree inventory, silviculture, ecology, wildlife, outdoor recreation and eco-tourism. Due to that, their landscape has changed, and many facilities had been developed to fulfill their establishment goals and objectives.

However, the question here is whether the development of those facilities is in harmony with the forest environment in terms of its identity and whether it is sustainable or not? It was argued that as much as the natural environment contained in the forests, the built environment influences the visitors' experience and impressions about how the management is fulfilling its mission of stewardship. This means, the landscape designs, especially the facilities in forest education centres must be fitted to forest setting, carefully constructed and meets the user's needs. This also shows that, landscape architectural identity of forest education centers is a very important focus as a way of showing respect towards the spectacular scenery and landscape settings, as well as to conserve natural resources. As a result, a good experience could be delivered to the user.

3. What is harmonized landscape design?

Harmonious landscape design can be referred to as the subordination of a structure to the environment and having buildings blended in with the landscape. The materials used, scale of the structure/building and form appears to fit into the existing landscape context. Materials used also need to reflect regional materials and be sympathetic with traditional forms and the existing landscape. It should also be a continuity of form, materials, colors and details among the structure/building within the existing area.

Harmonious landscape design is also closely related with identity. Nowadays, identity has been brought on the agenda of planners and designers to capture the concept of sustainability of place identity [10]. Here, place identity is strongly associated with the emotional type of attachment formed as a result of users' engagement and identification with places through activities and people that they have associated with [11]. Thus, the forest education centre's identity is reflected in the attributes of physical form (landscape architecture) and learning activity as well as the users' perception of the corresponding characteristics. It is also more than just the physical appearance but involves a "meaning" for the individual and the community [10]. At this point, researchers strongly believed that landscape architectural identity is also essential for the user's psychological existence and well-being which contributes to harmonious design. Norberg-Schulz [12] claimed that the architecture itself provided a physical attribute to space which facilitates habitation of the users as well as their mental and physical well-being. Therefore, landscape design elements in forest education centres must pay attention to the physical attributes of the forest setting to ensure it becomes an interesting outdoor education site.

Landscape design elements in forest education centres should consider a unique feature that highlights the forest identity [4]. The design should have a good combination between the natural environment and architectural aspects, for example, application of vernacular architecture [5]. On the other hand, the structure designs (e.g. pedestrian paths, signage's, shelters, etc.) as well as buildings must be visually in harmony with the surrounding natural environment [6]. Harmonious relationships of facilities with park landscapes should now include, in addition to concerns about form, color, and materials, along with concerns about issues such as landscape degradation, preservation of natural processes and protection of biological diversity [13]. On another aspect, Ceballos-Lascurain [8] and Walter [9] suggested that the structure and building designs in the natural setting for e.g. a recreational forest must highlight aspects of the surroundings. Therefore, the current development of our landscape design elements in forest education centers needs to be assessed, upgraded, developed or rehabilitated to improve their presentation of attractions.

Unfortunately, landscape design elements in forests are being built on an adhoc or in situ approach with the absence of a 'code of design' in creating a well-built environment of landscape and architectural quality [14]. This has contributed to inharmonious designs which led to the significant function of identity and meaning in securing the sense of place which was not adequately examined in the process. If this situation continues, it could raise an "abandon" syndrome which refers to unsafe guarding, unsustainability and conserving less of the surrounding natural and cultural heritage of the site. Whether or not this is realized, it could be observed that most of the landscape design elements in forest areas are quite similar with urban parks, for instance, the wakaf or shelter, pedestrian walkways and benches. This is true because most of their development are caught in the homogenizing forces of the mass media and are repeating the mediocrity of international fashion [15]. As a result, an identity crisis is displayed with its implications on architectural expression (for both building forms and landscape) resulting in harmonies design being neglected.

4. The need to harmonize landscape design for forest education centres

The need to have harmonized landscape designs for forest education centres would include the points below:

i. *Key to Sustainable Development Practices*. One of the primary aims of landscape design is to integrate human technology (buildings or other structures) with its natural environment. Its efforts are to achieve a better balance between the function and beauty, respect the character of the landscape and its sensitivity. Creation of environmental harmony through the objects provides a good relationship that brings balance and harmony to living beings, presents a quality of things agreeing, suitable and appropriate, as well as portraying the cultural value. Thus, applying harmony landscape designs has become one of the approaches to reach sustainable development for forest education centres.

ii. *To increase visual quality of the park*. Harmonize landscape design can increase the visual quality of the forest education centre because its design considers the buildings or structures to blend in with the landscape, applying appropriate colors and roof textures, foundation plantings, rough rock footings, and battered walls. What people see can influence how they feel and behave in the forest education centre. The user always expects to have quality experiences in the park through all the five senses, which is seeing, hearing, touching, smelling, and even tasting. Therefore, having buildings or structures blend in with the landscape is vital as it helps in increasing the visual quality of the park.

iii. *Provide appropriate education and enjoyment to user.* The aim of establishing forest education centres was to promote conservation of natural ecosystems, education focused on forest stewardship, and research related to forest ecology and management. Meanwhile, harmony landscape design objectives are to provide user and management facilities which are harmonious and visually pleasing in their simplicity and that, wherever possible, provide interpretation/information opportunities. Having harmonized landscape design with better planning of forest education centers will provide the users with a good learning environment and quality experiences. Furthermore, managers of forest education centres are often charged with protecting natural resources for future generations and providing appropriate user enjoyment.

iv. *Keys to guiding the selection of the most appropriate management responses.* Proactive strategies that embrace users and their perceptions towards forests should be developed to improve the health of park systems, wilderness areas and recreation destinations, including forest education centres. In this context, landscape design has a crucial role to play in achieving sustainability and to provide solutions for environmental problems. Therefore, understanding the need of harmonizing landscape design and their own impacts on the environment is a key to guiding the best building or structure design, as well as the most appropriate management responses. Furthermore, today's landscape design movement tends to address design problems.

v. *Approaches to conservation of cultural landscapes*. Landscape design has a significant impact on how users perceive and use the park. Thus, implementing harmony landscape design in forest education centres could provide a special character in which the values of the park are clarified and reinforced with cultural landscapes. Then, environmental education in forest education centres become more meaningful when it enables people to gain an understanding of how cultural values affect the environment. Cultural landscapes are a legacy for everyone. These special sites reveal aspects of our country's origins and development as well as our evolving relationships with the natural world. They provide scenic, economic, ecological, social, recreational, and educational opportunities helping communities to better understand themselves.

5. Methodology

The Sultan Idris Shah Forest Education Centre (SISFEC) was selected for a case study as it is one of the newest forest education centers in Malaysia located in Ayer Hitam Forest Reserve, Puchong, Selangor (**Figure 1**). Case studies were used because they are applicable to real-life, contemporary, human situations which allowed researchers to explore and investigate the contemporary real-life phenomenon through detailed contextual analysis of a limited number of events or conditions, and their relationships. Furthermore, it allowed the researchers to go beyond the quantitative statistical results and understand the behavioral conditions through the actor's perspective [17]. Yin [18] defined the case study research method "as an empirical inquiry that investigates a contemporary phenomenon within its real-life context; when the boundaries between the phenomenon and context are not clearly evident; and in which multiple sources of evidence are used."

To understand user's perception towards harmonies landscape design in forest education centers, data was gathered through a survey using a questionnaire with photographs as surrogates of the actual environment. A set of photos that show the landscape design elements in this center as attached to the questionnaire set, was

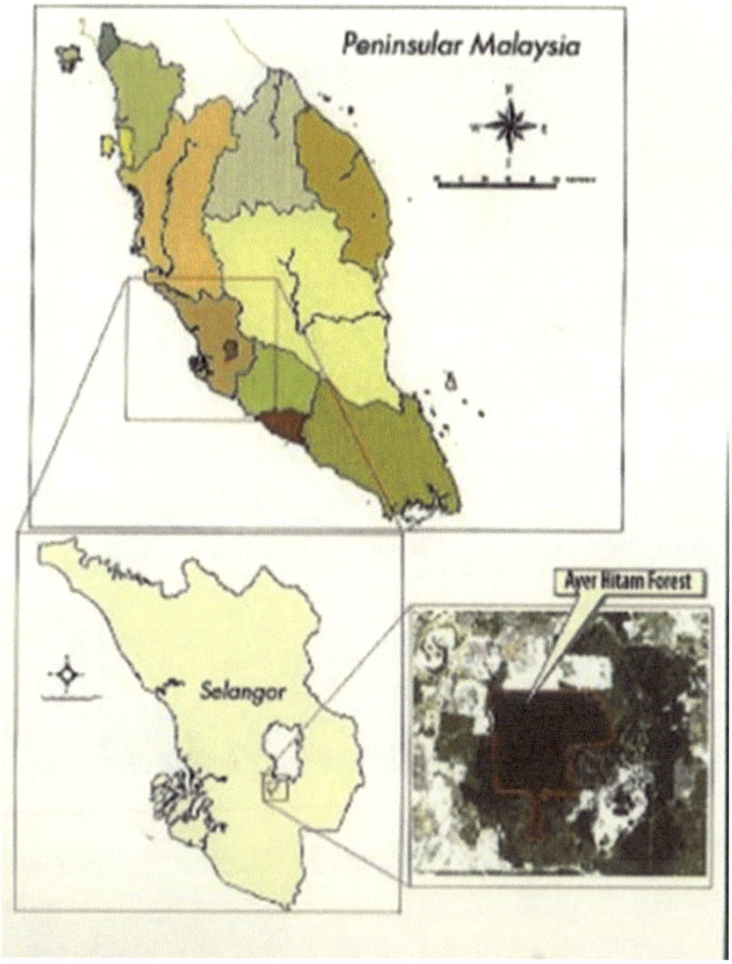

Figure 1.
Location of SISFEC at 3°00 53"N and 101°38'51"E [16].

prepared. The landscape design elements were divided into two categories (**Figure 2**). The first is the Buildings, which consist of the administration block, lecture hall, laboratory and chalets. Second includes Other Facilities consisting of a gazebo, walkway, boardwalk, outdoor toilets, guardhouse, signage, lighting, seating, railing and planter box. The landscape design elements are evaluated through the application of form, color, material, details and style/concept. These elements are the considerations that must be well thought out before the design work starts. It is obvious here to say that those elements were one of the factors that influenced perception. This is because forests have their own nature, and characteristics in it are an infinite variety of shape, colors and species living to gather in a perfect, logical, unquestionable way [19].

With the support of photographs, the respondents were questioned on the level of harmonizing for each of the elements based on four (4) Likert Scale. The scales are; 1 is not harmonize, 2 is slightly harmonize, 3 is harmonize and 4 is well harmonize. Each level of answer has the guided description for easier understanding (**Table 1**).

The survey was conducted in September 2019 until November 2019 and was basically carried out on weekends to target a higher number of respondents since

Figure 2.
Categories of landscape design elements used in survey (source: Authors, 2019).

Harmonize Aspect Likert Scale		Explanation
Form	1	Form does not harmonize with natural character of forest environment.
	2	Form slightly harmonizes with natural character of forest environment.
	3	Form harmonizes with natural character of forest environment.
	4	Forms are well harmonizing with natural character of forest environment and enhances the identity of SISFEC.
Color	1	Colors are too contrasted and does not harmonize with natural character of forest environment.
	2	Colors slightly harmonize with natural character of forest environment.
	3	Color is harmonized with natural character of forest environment.
	4	A color is well harmonizing with natural character of forest environment and enhances the identity of SISFEC.
Material	1	Material does not harmonize with natural character of forest environment
	2	Material slightly harmonize with natural character of forest environment
	3	Material used is harmonized with natural character of forest environment.
	4	Material used is well harmonize with natural character of forest environment and enhance the identity of SISFEC.

Harmonize Aspect Likert Scale		Explanation
Style/Concept	1	Building design's Style/Concept does not harmonize with natural character of forest environment.
	2	Building design's Style/Concept slightly harmonize with natural character of forest environment.
	3	Building design's style/Concept harmonizes with natural character of forest environment.
	4	Building design's style/Concept well harmonize with natural character of forest environment and enhance the identity of SISFEC.
Details	1	A detail does not harmonize with natural character of forest environment.
	2	Details slightly harmonize with natural character of forest environment.
	3	Details are harmonizing with natural character of forest environment
	4	Details are well harmonizing with natural character of forest environment and enhance the identity of SISFEC.
Overall Space Planning	1	Overall space planning does not respect the natural character of forest environment.
	2	Overall space planning slightly respects with natural character forest environment.
	3	Overall space planning respect with natural character of forest environment.
	4	Overall space planning respect the natural character of forest environment well and enhance the identity of SISFEC.
Summary of Landscape Architectural Image	1	Form, Color, Material, Details and Style/Concept of the Overall Architectural Image does not harmonize with natural character of forest environment.
	2	Form, Color, Material, Details and Style/Concept of the Overall Architectural Image are slightly harmonizing with natural character of forest environment.
	3	Form, Color, Material, Details and Style/Concept of the Overall Architectural Image are harmonizing with natural character of forest environment.
	4	Form, Color, Material, Details and Style/Concept of the Overall Architectural Image are harmonizing well with natural character of forest environment and enhance the identity of SISFEC

Table 1.
Likert scale of harmonize landscape design.

most education activities were held off during this period. A total of 150 respondents participated in the study. However, one respondent did not complete the survey form and was rejected; which resulted in 149 respondents only. In addition, respondents were selected based on onsite availability and their willingness to become participants. With help from the SISFEC staff, these groups of respondents were identified. Then, they were briefed on the procedure and were supplied with self-administered photo-questionnaires. The researchers were available to assist the respondents if they faced difficulties while answering the questionnaire.

The data gathered was analyzed using SPSS package version 24, using mean, frequency, percentage, Cronbach's alpha and reliability test.

6. Results and discussion

Questionnaire items were tested for reliability using Cronbach's alpha analysis to assess the internal consistency reliability of the questionnaire. According to Schmitt [20] and Spek et al., [21] an alpha of ≥0.7 was considered reliable. Therefore, the

Alpha Value for each variable used in the study is high in reliability standards and consistent with the value of 0.924 (**Table 2**).

Generally, respondents of this study perceived that the landscape design of SISFEC is slightly harmonized to harmonize with the mean score from 2.52 to 3.12 (**Table 3**). Only variable "Overall Space Planning" has a mean score of more than 3.00 which indicated the agreement on that design element has achieved the "harmonized" level with surrounding forest environment. The lowest mean scores with 2.52 are building's "form" and "material" where the respondents have high inclinations towards the harmonized level of design. Overall, respondents perceived the landscape and architectural image in SISFEC as harmonize with surrounding environment (mean score 2.74).

Meanwhile, **Table 4** shows in detail the frequency of each variable of respondents' perception on landscape design in SISFEC. Results had shown that 36.2% of respondents perceived the form of the buildings in SISFEC blend with the existing forest environment, whereas 14.1% of them thought the form was well blend with the forest environment. The respondents also perceived other facilities in SISFEC were blend (53.0%) and well blend (20.1%) with the forest. This had revealed that buildings and other facilities form in SISFEC was acceptable and in harmony with the existing environment. Form is one of the most important variables and has evocative effects on the way people perceive the surrounding as patterns. Thus, the best way to produce a harmonize design in the forest education center was the integration of park development with the landscape form, using native materials, textures, colors and respecting culturally significant resources. This study also observed that most of the building forms in SISFEC have a local architectural influence whereby Malay architecture is displayed on their roof style known as Bumbung Perak or 'Dutch style roof ridge' (**Figure 3**). This roof has been transformed into modern design where the slope of the roof is gentler, and the length of the eaves is also shorter. According to Riry [22], the application of

Variables	Alpha Value
Harmonize Aspect	
1. Buildings	
a. Form	0.915
b. Color	0.920
c. Material	0.920
d. Style / Concept	0.915
2. Building Details	
a. Details	0.914
3. Other Facilities	
a. Form	0.917
b. Color	0.919
c. Material	0.917
d. Style / Concept	0.913
4. Overall Space Planning	0.921
5. Summary of Landscape and Architectural Image	0.912

Table 2.
Alpha value for each variable.

traditional roof design in nature parks is suitable with the surrounding environment since the materials used are available locally such as wood and bamboo that enhances the harmonize design. Furthermore, simple traditional forms applied in

Variables	**Mean**	**Standard Deviation**
Harmonize Aspect		
1. Buildings		
a. Form	2.52	0.882
b. Color	2.54	0.904
c. Material	2.52	0.920
d. Style / Concept	2.65	0.961
2. Building Details		
a. Details	2.78	0.899
3. Other Facilities		
a. Form	2.87	0.799
b. Color	2.78	0.818
c. Material	2.86	0.811
d.Style / Concept	2.80	0.860
4. Overall Space Planning	3.12	2.74
5. Summary of Landscape and Architectural Image	2.74	0.823

Table 3.
Mean & standard deviation of landscape design.

Variables		**Frequency**	**Percentage (%)**
Haemonize Aspect	**Likert Scale**		
1. *Buildings*			
a. Form	1 not blend	18	12.1%
	2 slightly blend	56	37.6%
	3 blend	54	36.2%
	4 well blend	21	14.1%
b. Color	1 too contrast and not harmonize	21	14.1%
	2 slightly harmonize	47	31.5%
	3 harmonize	60	40.3%
	4 well harmonize	21	14.1%
c. Material	1 does not blend	21	14.1%
	2 slightly blend	52	34.9%
	3 blend	53	35.6%
	4 well blend	23	15.4%
d. Style / Concept	1 does not suite	22	14.8%
	2 slightly suite	37	24.8%
	3 suite	61	40.9%
	4 well suite	29	19.5%

Variables		Frequency	Percentage (%)
2. *Details*			
a. Details	1 does not blend	15	10%
	2 slightly blend	35	23.5%
	3 blend	67	45%
	4 well blend	32	21.5%
3. *Other Facilities*			
a. Form	1 not blend	9	6%
	2 slightly blend	31	20.8%
	3 blend	79	53%
	4 well blend	30	20.1%
b. Color	1 too contrast and not harmonize	9	6%
	2 slightly harmonize	41	27.5%
	3 harmonize	72	48.3%
	4 well harmonize	27	18.1%
c. Material	1 does not blend	8	5.4%
	2 slightly blend	35	23.5%
	3 blend	75	50.3%
	4 well blend	31	20.8%
d. Style / Concept	1 does not suite	11	7.3%
	2 slightly suite	38	25.5%
	3 suite	69	46.3%
	4 well suite	31	20.8%
4. *Overall Space Planning*			
Overall Space Planning	1 does not respect	5	3.4%
	2 slightly respect	26	17.4%
	3 respect	64	42.9%
	4 well respect	54	36.2%
5. *Summary*			
Summary of Landscape Architectural Image	1 does not harmonize	10	6.7%
	2 slightly harmonize	44	29.5%
	3 harmonize	69	46.3%
	4 well harmonize	26	17.4%

Table 4.
Frequency & Percentage of landscape design in SISFEC.

SISFEC allows the buildings to fit in with their neighbors and creates a balance with the existing environment.

In terms of color applied to buildings, 40.3% respondents perceived that the color was in harmony with the existing environment and 14.1% of them saw it as well harmonize with the surroundings. Respondents also perceived the color of

Figure 3.
A Malay traditional roof style of chalet in SISFEC (source: Authors, 2019).

other facilities was harmonizing (48.3%) and well harmonized (18.1%) with the forest setting. Here, the researchers understood that the respondents were pleased with the color used in the area since the color was not so different with the natural setting (**Figure 4**). Colors on existing park structures include brown, beige, and light pink. Overall, these colors are compatible with the colors found naturally in the surrounding environment. Bear in mind that color can create visual sensations other than emotional ones as well as can be used to simplify the forms and to break a building mass into smaller parts [23]. Therefore, the park management had given good consideration on the color effect of every element of a building, from the earthy colors of primary construction materials like wood, stone, brick, and marble, to the expansive variety of colors available for paint, doors, windows, siding, and trim. As a result, the colors are in harmony with the existing ambiance.

An analysis was conducted on materials, where the results are shown in **Table 4** that 35.6% respondents perceived that materials used is blend and 34.9% of them claimed the materials was slightly blend with the existing forest environment. They also agreed that materials of other facilities were blend (50.3%) and well blend (20.8%) with the forest condition. These results show that respondents have high inclinations towards the harmonized materials used in design. Building materials used in SISFEC include masonry block, stucco, wood, concrete, stone, and metal (on small features, such as railings) (**Figure 5**). The combination of materials used in existing building blends well with the natural environment. This is in line with the suggestions by Lippsmiler [24] and Schmid [5] who encouraged the use of local materials for construction works in protected areas, such as forests for sustainability purposes.

Figure 4.
Colors applied to the buildings and other facilities such as signage was harmonized with the existing environment (source: Authors, 2019).

Figure 5.
Building materials used in SISFEC include masonry block, concrete, stone and metal (source: Authors, 2019).

From the aspect of style or concept, SISFEC had applied a sustainable development or green technology concept. The architecture of the main complex was designed in modern forms that are innovative in construction technologies, particularly the use of glass, steel, and reinforced concrete (**Figure 6**). The administration building and lecturing halls were linked with boardwalks. There is also an observation tower at the Lecture Hall for students to view the forest's panoramic scenery surrounding the building. Each block was equipped with internet and Wi-Fi connections as well as CCTV for security purposes. The public can view tall and big trees around these buildings as well as small vegetation growth, untouched under the canopy. The researchers believe that this design approach was to harmonize the building with its physical surroundings, enhance community awareness, while reducing environmentally polluting substances to achieve a balance between man and the surrounding features. It is no wonder that 40.9% of the respondents saw that they are suited and 19.5% of them had perceived the style/concept of buildings in SISFEC was well suited with the forest environment, respectively. Respondents also perceived that other facilities such as boardwalk were also suited (46.3%) and well suited (20.8%) with the forest setting. It should also be stated here that SISFEC has implemented a green design approach where the building design tries to minimize the harmful effects of construction projects on human health and the environment [25].

In relation to details of building structure which includes the foundation, floors, walls, beams, columns, roof, stair, etc., this study found that almost half of the respondents (45.0%) perceived that the details part of buildings and other facilities in SISFEC blend in with the forest setting. Meanwhile, 21.5% of them claimed the details blend well with the forest environment. The other 23.5% of them saw it as slightly blending in with forest ambience. However, these figures had shown that

Figure 6.
Buildings in SISFEC had applied a green technology concept (source: Authors, 2019).

the respondents were more inclined towards harmonizing design where they made up a total up 90%. This is because they presume that the texture of bricks matches with the typical local area and the modern materials such as glass are to be used with care to maintain local character and achieve the aims of thermally efficient and sustainable building.

Furthermore, the footing system (**Figure** 7, right) was applied to buildings at certain parts and bridges (boardwalk) being built to connect each chalet in the lower areas that shows construction sensitivity. Additionally, retaining walls were constructed along the roads to protect them from soil erosion. This demonstrates that the management is very sensitive towards the natural environment and has minimized further disturbance. Indirectly, these efforts have created a harmonize environment and have protected the degradation of the existing landscape as well as enhancing the beauty of this area.

Finally, respondents were asked to rate the SISFEC space planning (**Figure 8**), whether the site has respected the existing environment or not. Results in **Table 4** had shown that 42.9% of the respondents perceived the space planning as giving

Figure 7.
Details of building and other facilities in SISFEC (Source: Authors, 2019).

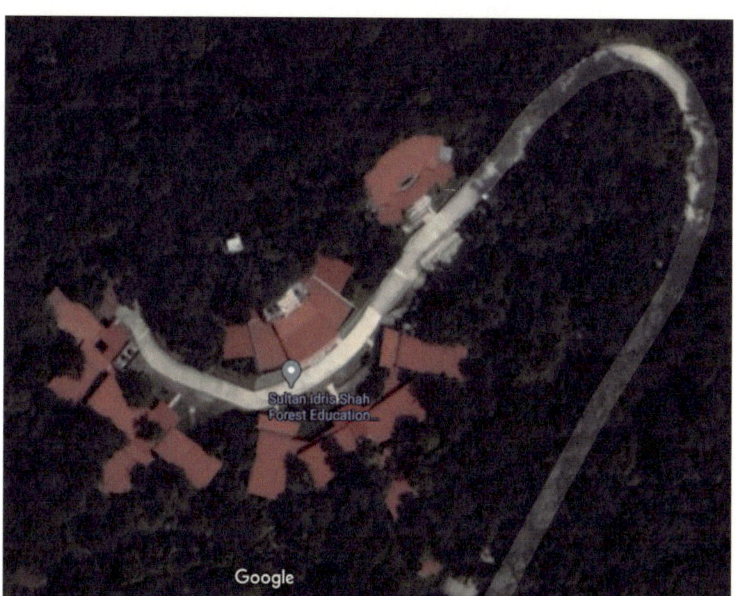

Figure 8.
SISFEC space planning (source: Google earth pro, retrieved on 5 mac, 2021).

respect to the existing environment. Meanwhile, 36.2% of them perceived it was well respect in the current environment. This is because the respondents saw the natural landscape has been conserved properly and kept in good condition. They noticed that tree cutting was strictly not allowed and the slope was kept in a natural setting and untouched, while major erosion was not found. The site still contains an untouched river landscape enclosed by the surrounding hills and have not suffered great disturbance. The most important thing is the buildings are not too huge and provides a contrast with the existing environment as well as the architectural features and details associated with local styles.

As conclusion to the findings, researchers understood here that harmonized landscape design is vital for Forest Education Centers to give users a closer look on nature and the architectural design to co-exist in mutual respect. Respondents of this study have expressed their perception when they rated the landscape design in SISFEC as harmonizes and inclined to well harmonize design. The summary of landscape architectural image in **Table 4** shows that 46.3% and 17.4% of the respondents perceived the design was harmonize and well harmonize, respectively. However, 29.5% of them claimed the design was slightly harmonized. It must be highlighted here that people give value to forest landscape that include esthetics, functional, and ethical values. Therefore, the park management should be aware of this and make an effort to improve and ensure all landscape design an element in SISFEC is in harmony with the forest environment.

Results of this study have emphasized that a good understanding of landscape design elements is necessary in creating a harmonize design. Each element is a crucial part of a visual message, and the combination of these has an impact on how the design is perceived. Thus, landscape design works should be made in agreement with the desires and demands of the tourists who spend their time in outdoor environments [26]. On another aspect, this study had shown that environmental harmony can be achieved when the aspect of consciousness and essence, aspect of energy and vitality and aspect of physical form being unites with each other [27]. Learning from the past as a source of possibility and inspiration for creating harmonious environments was also demonstrated as a good lesson.

This study also portrays that landscape design elements provide designers the ability to evaluate and deconstruct other designs as well as looking at a design from its basic, raw elements that would help a designer make certain decisions. Since forests are natural areas, therefore those who are involved in forest education centre's planning and development must perceive the site as a living being which leads towards environmental harmony. On this basis, it is necessary to consider the biological impacts of the negative effects of facility construction on the biological environment and their indirect impacts on the rational layout of land used in Forest Education Center planning, aiming to create a multi-directional interaction mechanism between ecological, economic, and cultural elements in this area [28].

7. Suggestions/recommendations

Designing a Forest Education Centre is a true challenge on the application of harmonizing design approach which considers the relationship between man-made structures and the surrounding natural environment. A harmonize landscape design approach has been applied since early human civilization. It shows interdependently strong the force was between nature and the need for human habitation. Through the test of time, the design approach had evolved in the theoretical framework, but the goal remains in finding commonality between architecture and natural environment.

It was suggested here that the development of forest education center in Malaysia should use a green building materials in construction activity such as lime, sand-lime brick, eco-friendly tiles and colored plaster, which are more effective than the conventional materials. Those said materials can be reused, renewed or recycled, sustainably harvested, rapidly renewable, non-toxic, and local without adversely polluting the environment. The structures or buildings should be painted using eco-friendly paint or low volatile organic compound coatings to eliminate potential health hazards and improve indoor living environments.

Some harmonize design approach in architecture that can be applied for Forest Education Centers are Eco-architecture, Organic Architecture, Vernacular Architecture and Green Building. These design approaches are the labels of architectural style with the same objective to achieve a harmonize design goal. The harmonize design approach play a vital role in establishing a Forest Education Center for the function of environmental education and awareness in an Educational Forest as well as to minimize the conflict between forests, urbanization and the environment for a sustainable development in the communities, region and the world. The management of Forest Education Centers can choose one of the proposed approaches depending on the site suitability. Additionally, a combination of the proposed approaches also could be applied depending on site suitability.

i. *Eco-Architecture*

Eco-Architecture means the application of ecological principal to architecture, typically in the design of building which promotes environmental conservation that harmonizes with their natural surroundings. An example of Eco-Architecture in Malaysia is the Menara Mesiniaga building, designed by Ken Yeang (**Figure 9**). The building was designed according to the local climate and seasons which integrated with the natural landscape. The building was highly recognized globally and received the Aga Khan Award in 1995.

Moreover, eco-architecture seeks to minimize the negative environmental impacts of structure through improved efficiency and the use of sustainable construction materials. As a result, it reduces the consumption of fossil fuels while providing esthetically pleasing structures that are good for the environment. In order to implement eco-architecture in forest environments, it is crucial to understand the key elements of Eco-Architecture which include site context, nature connection, natural processes, environmental impact and people (refer **Table 5**).

Obviously, Eco-Architecture pays attention to the ecological aspects during the planning stage which is soil suitability, site selection, water resources, and waste management. It also utilized the natural elements as to which one is biodegradable, renewable, and clean elements with low-embodied energy for building construction. Eco-architecture is a sustainable architecture that involves a combination of values which are esthetic environmental, social, political, and moral. There is a need to engage in the central aspect of practice, designing and building in harmony with the environment. It needs to rationally think about a combination of issues, including sustainability, durability, longevity, appropriate materials, and sense of place. Thus, Eco-architecture is always one of the best harmonize approaches to maintain the sensitivity of the forest spirit that is suitable to be applied for Forest Education Centers.

Figure 9.
Eco-architecture of Menara Mesiniaga, Selangor, Malaysia (source: AD classics: Menara Mesiniaga/T. R. Hamzah & Yeang Sdn. Bhd (2015). Retrieved form https://www.archdaily.com/774098/ad-classics-menara-mesiniaga-t-r-hamzah-and-yeang-sdn-bhd).

ii. *Organic architecture*

Organic architecture is a philosophy of architecture which emphasizes on the harmony between human habitation and the natural setting. This design approach is also known as democratic and individual and which becomes the factor when one creates a building with the consideration of the environment, the identity, the lifestyle and the uniqueness of the place. Martin Howard & Simon Forty [29] claimed that organic architecture, however, meant that the architect was not only involved in the structure of the building, but the contents as well.

A good example of Organic Architecture is Falling water, designed by Frank Lloyd Wright in 1865–1959 (**Figure 10**) [30]. Falling water had been built at the waterfall and steep slope resulting in a very dramatic view of harmonize composition between natural landscape and manmade structure in the rural area. The design emphasizes on the horizontal striations of stone masonry with daring cantilevers of colored beige concrete that blend with native rock outcroppings and the wooded environment. As Frank Lloyd Wright [30] said:

"*A building should appear to grow easily from its site and be shaped to harmonize with its surrounding if nature is manifest there*".

1	Understanding context	Understanding the site contextual with the forest environment is vital before starting any design in a forest. Forests are highly sensitive areas in all aspects, especially on the ecological systems found in the forest. By emphasizing the contextually of the site, it will produce an efficient design without damaging the forest environment. Among the factors to be considered in understanding the context of a forest site are geographical location, temperature, plant and animal habitat, water source, high viewpoint, forest canopy and so on.
2	Connecting with nature	Every building design should have a close relationship with the forest's original state. This is because the design of the building will affect the forest environment. The perfect and beautiful combinations of building structures and natural environment will create a harmonize relationship s with the forest.
3	Understanding natural process	Understanding natural processes in the forests plays an important role to the natural habitat. This is because life in the forest is dependent on each other's ecosystem. Consequently, the impact of the designs in the forest will include the organism's response to their environment and how the organisms are adapting towards the presence of strangers in their habitats.
4	Understanding environmental impact	Environment-friendly designs will have a good impact on the forest environment, the original habitat of plants and animals as well as human beings as a consumer. Careful design planning and the use of natural elements will reduce pollution, as well as conserve and preserve forest areas at it best. This can also reduce the harmful effects towards forest biodiversity.
5	Understanding people as a user	Another factor that is affecting the result of design are humans as the main user. Every space that is designed should consider the individual's needs. After understanding the type of user only then can it be identified according to their needs and types of design. As such, interactions between humans and nature can be harmoniously created.

Table 5.
Key element of eco-architecture.

Figure 10.
Falling water by frank Lyod Wright (source: Falling waters (2017). Retrieved form https://iknowexpo.org/blog/travel-opportunities-seniors/attachment/falling-waters/).

The key elements of Organic Architecture are simplicity, style, sympathy with the environment and materials used. **Table 6** below summarizes the key elements of Organic Architecture.

It can be understood here that Organic Architecture has a significant relationship between nature, and its understanding and appreciation, the

1	Simplicity	Simplicity is one of the most important design principles in designing a building in a forest. This is due to the natural and peaceful environment of the forest that makes the concept of simplicity more appropriate. Similarly, the use of finishes that is not overwhelming and the inconceivable design will make the design more harmonize.
2	Style	The style of design is very important to enhance the ideas. Even though it is in the forest, it is still considered important for the purpose of aesthetical and comfort. In addition, the suitability of the style should be harmonizing with the forest environment as well as the local culture. The remarkable style will produce a strong sense of place.
3	Sympathy with the environment	Natural environment with flora and fauna is the key consideration prior to the planning and design works in the forest. Being sympathetic with the forest environment can create a harmonious design and produce effective interactions between building structures, humans and nature. In-depth understanding and mastery of forest situations can make individuals understand the principles of life more and them to become closer to natural environment.
4	The "nature "of materials	The selection of suitable building materials that corresponds to local conditions indicates the simplicity of building structures in the forest. It is more relevant to the forest's sensitivity due to its existence in living things and its unique view while maintaining the natural concept of forests. Minimizing the selection of materials that will be used is more practical than to diversify them and to harmonize with nature.

Table 6.
Key elements of organic architecture.

use of horizontal expression, logical design, plus appropriate scale and equality in the construction of architectural components. Furthermore, the design style of organic architecture that is characterized by the suitability of the use of materials and structures will shape the identity of an Educational Forest. Hence, Organic Architecture was believed to also fit into the Forest Education Centers where the design that emphasizes simplicity, humility, and respect of nature will benefit the visitors, researchers and the management, as well as enduring in the forest environment.

iii. *Vernacular Architecture*

Vernacular architecture is an approach of design of a building or structure by the skills and expertise of local people without any formal training in design. The design is based on local needs based on a specific time and place which is not replicated from elsewhere. The expertise has been guided by a series of conventions built up in their locality where the consideration of functionality is the dominant factor and the use of local materials is the primary concern. The design tends to evolve over time to respond to the environmental, cultural, spiritual, technological, economical and historical context to become more refined. It is also synonymous with primitive, nomadic or traditional architecture, ethnic architecture and aboriginal architecture.

Kırbaş & Hızlı [31] emphasized that Vernacular Architecture has been an inspiration for innovations in environmental and socio-economically sustainable design and planning. The example of Vernacular Architecture is the traditional Batak house in Sumatra, Indonesia (**Figure 11**), where the use of bamboo and woods with the thatched roofs are commonly practiced. Other materials which are also frequently used depending on the locality

Figure 11.
A traditional Batak house, Sumatra, Indonesia (source: Traditional Batak house (2019). Retrieved from https://www.britannica.com/topic/Batak/media/55738/119035).

include clay, wood, animal's skin and others. All those examples have been shown in traditional housing, where the intended climatic and environmental solutions within sustainable design have already been achieved by local implementations.

Vernacular Architecture approach for a Forest Education Center would involve a deep respect and perfect communion with the natural forest environment as well as the perfect relation and understanding of the user's needs. The abundance of forest species may be used as a material with sustainable quantity to make sure it still respects the conservation needs of the forest. The unique local cultural characteristic could be harmonizing with the Forest Education Centre's design language, so the identity of the place will be created in a more meaningful manner. Apart from that, Forest Education Centres also need to take into consideration micro-climatic factors such as lowland dipterocarp forest's setting which has different forest sensitivity from peat-swamp forest or montane forest.

The key elements in deliberating Vernacular Architecture are the climatic influences, cultural significant, environment, construction elements and materials (refer **Table 7**).

In summary, the Vernacular Architecture approach in designing Forest Education Centers will be a large repository of natural and cultural heritage that illustrates a genuine and symbiotic relationship with spirits of the forest. Consequently, the relationship which is mediated through knowledge and values can contribute to the creating of harmonious design in the Forest Education Center itself.

iv. *Green Building*

Green building is a building that reduces or eliminates negative impacts, and can create positive impacts, on our climate and natural environment. It

1	Climate	The influence of climate on vernacular architecture is one of the important factors that affects the overall planning and design. Differences in temperature, light, humidity and wind movement at a place determine the materials to be used in designs, particularly within the forest area. For example, emphasizing on the use of wood is well suited to the tropical climate, furthermore with proper handling and installation, the material will be longer lasting due to its compatibility with climate and local weather. The result of this method makes a building stand up for several hundred years.
2	Culture	Sensitivity to culture as well as understanding the pattern of community life of a place is the key foundation in producing a design that is closer to the way of life of the society. There are similarities and differences in indigenous groups in terms of customs, way of life, interaction patterns that will influence significantly to the size and usage of spaces in Vernacular Architecture. Cultural aspect plays an important role in meeting the objectives and to ensure the successfulness of the design.
3	Environment, construction elements and materials	Natural material found around the local area are the main selection for vernacular building sources. For example, forest-rich wood or bamboo areas will make use of it as a main building material and consequently it affects the style of vernacular architecture. This clearly shows that vernacular architecture is cost effective, quality, affordable, beautiful and durable.

Table 7.
Key elements of vernacular architecture.

is the set of practice of constructing buildings or structures and using processes that are environmentally responsible and resource-efficient throughout a building's lifecycle from planning to design, construction, operation, maintenance, renovation and deconstruction. According to Samer [32], the "Green Building" is an interdisciplinary theme, where the green building concept includes a multitude of elements, components and procedures which diverges to several subtopics that are intertwined to form the green building concept. Ding & Zhikun [33] further emphasized that the adverse impact of construction on the environment significantly promotes the development of the green building concept worldwide.

In order to achieve a Green Building's standard, there are several factors that should be considered which includes efficient use of energy, use of renewable energy such as solar, waste reduction measures, enabling of recycling and use of non-toxic materials. It is the consideration of the quality of life of occupants with a design that enables adaptation to a changing environment as well as the quality of the environment in design, construction and operation. The example of Green Building (**Figure 12**) shows the detail features in the building.

Table 8 below lists the key elements in deliberating Green Building which includes resources efficient, in the area of energy; healthy for both physically and psychologically; feeling comfortable, responsive and flexible; and ecologically integrated system of impacts based.

Instantaneously, Green Building is a sustainable or high-performance building. The functionality of Forest Education Centers with the approach to green buildings can improve the harmonious aspects with the surrounding forest environment, users can enjoy healthier atmospheres, free of unnecessary pollution and waste. This approach will provide an excellent impact through the efficient design of the building which can complement the objective of sustainable educational forest

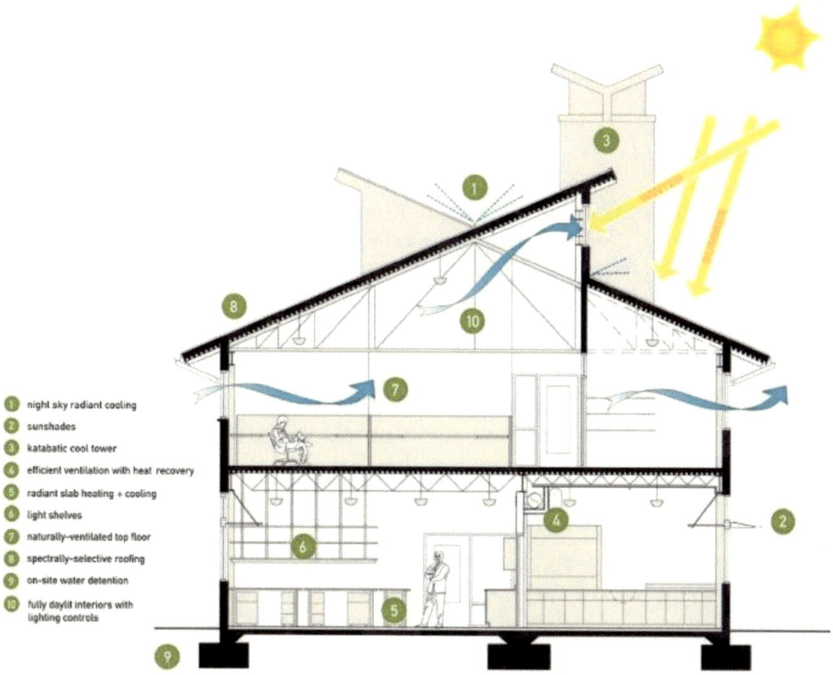

Figure 12.
Features of a green building (source: 14 green building sustainable design images (2012). Retrieved from http://www.newdesignfile.com/post_green-building-sustainable-design_184941/).

1	Resources efficient, in the area of energy	Green building responds to the environmental needs which certainly leads to a better quality of human life. Determining areas that provide a lot of natural energy such as light, water, and wind can make a green building work efficiently in accordance to the environmentally friendly design approach.
2	Healthy, both physically and psychologically	Normally, humans will spend a lot of time in the building as shelters for daily activities. If the built-in design does not consider the user's health, this will make the productivity decrease due to the impact of the deterioration of the user's health while inside the building. Therefore, the application of the green building concept in a design will have a positive impact on both the user and the environment.
3	Comfortable, responsive and flexible	Rationally, the safety and comfort aspects of being in a building are important, besides the basic facilities provided to the user. Evidently, the design of green buildings can provide a comfortable and clean environment without any pollution, especially air and water. The diversity of natural resources' usage in buildings can provide benefits and continuous advantages. Consumers are also free to conduct daily activities with a healthy and positive aura.
4	Ecologically based, particularly as an integrated system of impacts	Basically, this approach also signifies environmental concerns and is aligned with sustainable development concepts. Preserving ecological systems in the forest, for example, should be sustained as not to cause further damage involving the entire inhabitants of the earth. The harmony between buildings design and nature is a major factor in the ecological design.

Table 8.
Key elements of green building.

awareness through research and educational activities. Therefore, this kind of approach would be suitable for forest education centers environment to consider.

8. Conclusion

Forests are significant in our lives and other living things such as animals (e.g. snakes, insects, birds, lions, leopard, monkeys, etc.). Meanwhile, a healthy forest environment is critical for human and animal lives since improper management of this environment would seriously affect humans and animals lives. One of the ways to protect it is by educating people on how to love and manage forests in a proper way.

This study has highlighted that the development of Forest Education Centers must be harmonizing with the forest environment for its sustainability. People will accept these kinds of development when they believed that the development was harmonize with the surrounding and does not give big impact to the environment and their lives. Without proper development planning, it will affect the balance of nature, such as soil, water, air and climate, plants and animals, diversity, characteristic features and beauty and the recreational value of nature as well as the existing and foreseeable land uses. Therefore, harmonize landscape design needs to be prepared for the areas concerned to maintain and regain ecological integrity and enhance humans well-being towards sustainability. It is also necessary to ensure that the proposed landscape elements harmonize with the existing forest environment.

Hence, the forest education center development must be planned before implementation to make sure natural landscape resources (e.g. forests, rivers, hills, waterfalls, stones, etc.) are continuously protected and preserved. It is also to ensure that human-made landscape elements can harmonize with the existing environment. Thus, facilities/infrastructures must conform to environment-friendly, low-impact architecture; renewable including solar energy, waste recycling, rainwater harvesting, natural cross-ventilation, no use of asbestos, controlled sewage disposal, and merging with the surrounding landscape. The design should enhance local architecture, use local materials as well as portraying local art and culture.

Particular attention should be given to local landscape and building traditions, boundary treatments, mix of materials, scale and proportion. New buildings should fit in and make a positive contribution to their surroundings. Where no overriding context exists, efforts should be made to create a positive and distinctive character for the locality with roots in the local character of the District. The basis for any development should be to combine concerns for both sustainability and local distinctiveness.

While seemingly simple to prepare a comprehensive harmonize landscape design and development plan, it remains a challenge because it requires commitment and allocation. We must think ahead for sustainable development in the long run that will provide benefits to all parties, despite the inevitable challenges, which should come first.

Acknowledgements

This study is sponsored by Universiti Putra Malaysia under the Geran Putra 2018 (GP/2018/9652800).

Author details

Mohd Kher Bin Hussein* and Nor Syuriaty Bt Jaafar
Department of Landscape Architecture, Faculty of Design and Architecture, Universiti Putra Malaysia, Selangor, Malaysia

*Address all correspondence to: mohdkher@upm.edu.my

References

[1] Izaidah, Talib. (2015). Overview of Forestry Sector in Peninsular Malaysia. *International Journal of Science*. Volume 4: December 2015 (12). Published at: http://www.ijsciences.com/pub/issue/2015-12/ DOI: 10.18483/ijSci.882; Online ISSN: 2305-3925; Print ISSN: 2410–4477

[2] NRE. 2011. Malaysia Second National Communication to the UNFCCC. Ministry of Natural Resources and Environment. Putrajaya, Malaysia.

[3] Forest Department of Peninsular Malaysia (2012). Compendium of amenity forests and state parks forests in Peninsular Malaysia. Retrieved from www.forestry.gov.my

[4] Jacobs, A.B. (1993). Great Streets. (Massachusetts, MIT Pres).

[5] Schmid, P. (1983). Bio-logis Architectur, 2nd ed. (Koln-Braunsfeld: Rudolf Muller).

[6] Schneider, R. (1981). Natural Building. (Stuttgart: Institut fur leichten Flachen tragwerke).

[7] Doczi, G. (1994). The Power of Limits - Proportional harmonies in nature, art and architecture. Boston, London: Shambala.

[8] Ceballos-Lascurain, H. (1996). Tourism, ecotourism and protected areas - The state of nature-based tourism around the world and guidelines for its development. IUCN, Gland, Switzerland, and Cambridge, UK.

[9] Walter, G. (1987). Design in Architecture. (Koln: Rudolf muler).

[10] Isil Kaymaz (2013). Urban landscape and identity, advances in landscape architecture. Dr. Murat Ozyavuz (Ed.), ISBN: 978–953–51-1167-2, InTech. DOI: 10.5772/55754. Available from: http://www.intechopen.com/books/advances-in-landscape-architecture/urban-landscape-and-identity

[11] Noorsidah Ujang & Kamariah Dola. (2007). Linking activity and place attachment dimensions in enhancing the sense of place. *Alam Cipta. International Journal on Sustainable Tropical Design Research and Practice*. Vol.2 (1): 59–67.

[12] Norberg-Schuls, C. (1985). The concept ow dwelling on the way to figurative architecture. New York: Rizolli.

[13] Zube, E.H. (1993). The Search for Harmony in Park Development. Visual Quality of Built Environments in National Parks. Department of Geography and Regional Development, University of Arizona, Tucson,US.

[14] Noorizan Mohamed. (1995). Conservation in Malaysia: landscape, tourism and culture. Unpublished PhD. Thesis. University of York, UK.

[15] Noor Azizi, A. (2003). Landscape architecture in Malaysia-in search of identity. *Alam Cipta*. Fakulti Rekabentuk dan Senibina. (Universiti Putra Malaysia).

[16] Faridah-Hanum. (2009). Ayer Hitam Forest. The Green Lung of Klang Valley. Serdang, Selangor: Penerbit Universiti Putra Malaysia.

[17] Zaidah Zainal. (2007). Case study as a research method. Jurnal Kemanusiaan, 9 (Jun 2007): 1–6.

[18] Yin, R.K., (1984). Case Study Research: Design and Methods. Beverly Hills, Calif: Sage Publications.

[19] Anthony Peter, and Oladipo Francis Olukayode. (2017). "Architectural Design Harmonising With The Surrounding Features." International

Journal of Research - Granthaalayah, 5 (11): 102–108. https://doi.org/10.5281/zenodo.1069176.

[20] Schmitt N (1996). Uses and abuses of coefficient alpha. Psychol Assess, 8, 350–3.

[21] Spek V, Lemmens F, Chatrou M, et al (2013). Development of a smoking abstinence self-efficacy questionnaire. Int J Behav Med, 20, 444–9.

[22] Riry Zaimora. (2006). *Pelaksanaan piawaian pembangunan fasiliti penginapan dalam kawasan hutan lindungan di Malaysia*. MSc. Thesis. Universiti Teknologi Malaysia, Skudai, Malaysia.

[23] Bell, S. (1993). Elements of Visual Design in the Landscape. E & FN Spon, London.

[24] Lippsmiler, G. (1997). Bangunan Tropis 2nd. Ed.: Jakarta. Air Langga.

[25] Roy Madhumita. (2008). Importance of green architecture today. Dept. of Architecture, Jadavpur University, Kolkata, India. Retrieved from https://sari-energy.org › SAWIE › SAWIE_11–2008 › SAWIE_11–2008 at 25.07.2019.

[26] Dönmez, Y. and Türkmen, F. (2018). The relation between the landscape design and brand image in purchase preferences of tourists: the case of Safranbolu and Nevşehir, in Turkey. Applied Ecology And Environmental Research 16(1):629–643. DOI: http://dx.doi.org/10.15666/aeer/1601_629643.

[27] Ardita Byci Jakupi. (2016). Landscape Architecture as Environmental Harmony. International Journal of Contemporary Architecture" The New ARCH"Vol. 3, No. 3 (2016).

[28] Yu, X. H., Wang, X., Ren, Y. G. & Liu, J. C. (2019). The landscape evaluation system of ecotourism villages in Qinling Mountains. Applied ecology and environmental research 17(4):8955–8968. DOI: http://dx.doi.org/10.15666/aeer/1704_89558968.

[29] Martin. H. & Simon. F. (1999). Frank Lloyd Wright: A Visual Encyclopaedia. London: PRC Publishing Ltd.

[30] Frank Lloyd Wright (1925). "Frank Lloyd Wright: The Complete 1925 "Wendingen" Series", p.11, Courier Corporation

[31] Kırbaş, B., & Hızlı, N. (2016). Learning from Vernacular Architecture: Ecological Solutions in Traditional Erzurum Houses. *Procedia - Social and Behavioral Sciences, 216*, 788–799. 10.1016/J.SBSPRO.2015.12.076

[32] Samer, M. 2013. Towards the implementation of the Green Building concept in agricultural buildings: a literature review. Agric Eng Int: CIGR Journal, 15(2): 25–46. Available from: https://www.researchgate.net/publication/245543328_[accessed Jan 18 2021].

[33] Ding, Z et al., (2018). Green building evaluation system implementation. *Building and Environment, 133,* 32–40. 10.1016/J.BUILDENV.2018.02.012